Modern Communications

Designed for a single-semester course, this concise and approachable text covers all of the essential concepts needed to understand modern communications systems. Balancing theory with practical implementation, it presents key ideas as a chain of functions for a transmitter and receiver, covering topics such as amplification, up- and down-conversion, modulation, dispersive channel compensation, error correcting codes, acquisition, multiple-antenna and multiple-input multiple-output antenna techniques, and higher-level communications functions. Analog modulations are also presented, and all of the basic and advanced mathematics, statistics, and Fourier theory needed to understand the concepts covered are included. Supported online with PowerPoint slides, a solutions manual, and additional MATLAB®-based simulation problems, it is ideal for a first course in communications for senior undergraduate and graduate students.

Daniel W. Bliss is a professor in the School of Electrical, Computer, and Energy Engineering at Arizona State University (ASU), and is the Director of ASU's Center for Wireless Information Systems and Computational Architectures. He is a co-author of *Adaptive Wireless Communications: MIMO Channels and Networks* (Cambridge University Press, 2013).

"Strikes an elegant balance between fundamental concepts, their applications, and the depth of explanation. It's the kind of book that you want to hand to all beginners in wireless."
Ashutosh Sabharwal, Rice University

"An ideal introduction to modern communications systems. Theoretical principles and practical considerations are presented in an integrated fashion, and the material is introduced in an intuitive manner with a logical progression of ideas, making this the perfect text for a beginner with an interest in pursuing serious study of modern communications systems. This book should be on the required or recommended text list of all introductory communications courses."
Siddhartan Govindasamy, Boston College

Modern Communications

A Systematic Introduction

DANIEL W. BLISS

Arizona State University

CAMBRIDGE
UNIVERSITY PRESS

University Printing House, Cambridge CB2 8BS, United Kingdom

One Liberty Plaza, 20th Floor, New York, NY 10006, USA

477 Williamstown Road, Port Melbourne, VIC 3207, Australia

314–321, 3rd Floor, Plot 3, Splendor Forum, Jasola District Centre, New Delhi – 110025, India

103 Penang Road, #05–06/07, Visioncrest Commercial, Singapore 238467

Cambridge University Press is part of the University of Cambridge.

It furthers the University's mission by disseminating knowledge in the pursuit of education, learning, and research at the highest international levels of excellence.

www.cambridge.org
Information on this title: www.cambridge.org/9781108833431
DOI: 10.1017/9781108980685

© Cambridge University Press 2022

First published 2022

Printed in the United Kingdom by TJ Books Limited, Padstow Cornwall, 2022

A catalogue record for this publication is available from the British Library.

Library of Congress Cataloging-in-Publication Data
Names: Bliss, Daniel W., 1966– author.
Title: Modern communications : a systematic introduction / Daniel W. Bliss.
Description: Cambridge ; New York, NY : Cambridge University Press, 2022. |
    Includes bibliographical references and index.
Identifiers: LCCN 2021012635 | ISBN 9781108833431 (hardback)
Subjects: LCSH: Wireless communication systems.
Classification: LCC TK5103.2 .B56 2022 | DDC 621.384–dc23
LC record available at https://lccn.loc.gov/2021012635

ISBN 978-1-108-83343-1 Hardback

Additional resources for this publication at www.cambridge.org/blisscomms

Contents

Preface

It is a very sad thing that nowadays there is so little useless information.

Oscar Wilde

In writing this text, my aim is to provide a systematic and efficient introduction to the way in which modern communications systems are designed and analyzed. While there are many wonderful undergraduate texts on this topic (for example, [1–3]) that commonly employ a historically motivated structure, this textbook provides an alternative that is hopefully more efficient. I consciously chose to keep this text relatively short to encourage a linear reading of the entire text. I focus on wireless communications, but the ideas presented are applicable to other types of communications systems. This textbook will allow you to teach all the critical topics in a semester-length course.

The content of this textbook addresses the topics necessary to understand issues that are critical to building real-world communications systems. This focus differentiates my text from some that miss important practical topics while overemphasizing techniques that are now less important. The textbook is split into two parts. The first part is the essential portion of the text. The second part contains useful background information. I use notional transmitter and receiver radio chains to guide this introduction. I introduce the idea of channel capacity early in the textbook. This allows the reader to have a sense of the limits of performance that motivate the engineering choices we make. Additionally, I stress the effects of hardware limitations, such as noise figure, phase noise, analog-to-digital converter (ADC) and digital-to-analog converter (DAC) quantization, and linearity. I emphasize up- and down-conversion approaches that are currently used, such as direct conversion and digital intermediate frequency (IF). I provide a short introduction to forward error correction and draw the connection to capacity. I discuss the critical topics of acquisition and synchronization that are often overlooked. I discuss dispersive channels and emphasize approaches such as orthogonal frequency-division multiplexing (OFDM) and adaptive equalization to address these channels. I also introduce multiple-input multiple-output (MIMO) communications, and with an eye toward future systems I consider adaptive multiple-antenna processing. Finally, I review or introduce the mathematical tools, such as linear algebra and statistics, to perform the calculations required to understand communications systems.

Specifically, I organize this textbook into the following topics. In Chapter 1, "Notation," I provide a quick overview of the mathematical notation used. In Chapter 2, "Basic Radio," I provide a quick survey of concepts used in communications systems, focusing on wireless radio systems. These concepts include signal sources, channels (for example, the Friis equation for line-of-sight propagation), noise sources, signal-to-noise ratio, theoretical limits, nonidealities, antennas, waveforms, and representative transmit and receive processing chains. In Chapter 3, "Fundamental Limits on Communications," I introduce the concepts of channel capacity and source compression, including the idea that there is a fundamental achievable bound on data rate (Shannon's channel capacity) with a sketch of a proof of the bound. I also discuss how to use the Shannon limit to help in system design. In Chapter 4, "Amplifiers and Noise," I present an overview of amplifiers and sources of noise, including an overview of low-noise amplifiers (LNAs) with a derivation of the Friis formula. I discuss nonidealities of amplifiers including nonlinearity with an explanation of the third-order input intercept point (IIP3). In Chapter 5, "Up- and Down-Conversion," I introduce techniques to move signals between baseband and carrier frequencies, including multiple up- and down-conversion techniques. I connect the mathematical formulations with the basic block diagrams of various direct conversion and superheterodyne approaches. In this context, I discuss DACs and ADCs with a corresponding discussion of dynamic range. I also introduce how the combination of crystal oscillators and phase-locked loops (PLLs) are used to construct frequency synthesizers that are integral to the operation of up- and down-conversion techniques. In Chapter 6, "Modulation and Demodulation," I present modulation and demodulation techniques to move between bits and in-phase and quadrature baseband voltages. I discuss various constellation approaches, including binary-phase-shift keying (BPSK), quadrature-phase-shift keying (QPSK), 8-phase-shift keying (8-PSK), 16-quadrature-amplitude modulation (16-QAM), and differential approaches. I compare hard decision and soft decision demodulation approaches. I also sketch the derivation of the modulation-specific capacity result. I provide a discussion of the relationship between pulse-shaping filters and spectral shapes. In Chapter 7, "Dispersive Channels," I introduce the concept of propagation channels with delay spread and provide receiver and waveform techniques to compensate for these effects, including adaptive equalizers and OFDM. For OFDM, I motivate the idea of a cyclic prefix and discuss subcarrier spacing design. In Chapter 8, "Error Correcting Codes," I present forward error correction approaches at an introductory level. While a proper treatment would take a semester, I do provide an overview of systematic linear block codes (Hamming codes, for example). I discuss generator matrices, parity check matrices, and syndromes. I also provide an introduction to convolutional codes with an explanation of Viterbi demodulation. In Chapter 9, "Acquisition and Synchronization," I survey techniques for acquiring a communications link and synchronizing between radios. I introduce various acquisition and synchronization approaches such as energy, cross-correlation, and autocorrelation detector techniques. I discuss the receiver operating characteristic (ROC) curves for these approaches. I also derive the maximum likelihood

(ML) techniques for spectral and temporal synchronization. In Chapter 10, "Radio Duplex, Access, and Networks," I provide an introduction to how radios inter-act in terms of use of the channel. I introduce the ideas of time-division duplex (TDD) and frequency-division duplex (FDD) approaches. I also introduce vari-ous access control approaches, including Aloha and carrier-sense multiple access (CSMA), time-division multiple access (TDMA), frequency-division multiple access (FDMA), and code-division multiple access (CDMA). I provide simple examples of CDMA spreading and de-spreading. In Chapter 11, "Multiple-Antenna and Multiple-Input Multiple-Output Communications," I introduce theory and techniques for multiple-antenna and MIMO techniques. I introduce the concept of multiple-antenna receivers and MIMO communications systems, as well as spatially adaptive beam-forming. I also discuss MIMO channel capacity and some simple space–time coding approaches. In Chapter 12, "Analog Radio Systems," I review analog radio tech-niques for historical completeness. I review amplitude and phase modulations and various carrier and sideband suppression approaches. To provide a mathematical background, I discuss the Hartley modulator and the associated Hilbert transform. I provide an analytical presentation of the spectrum of a sinusoidal phase modulation. Finally, I provide an analysis of how to use PLLs to demodulate angle-modulation techniques.

In Part II, "Mathematical Background," I review and introduce concepts that are useful in understanding communications systems. In Chapter 13, "Useful Mathe-matics," I provide a range of useful engineering mathematics, including complex variables, linear algebra, multivariate calculus, and special functions. In Chapter 14, "Probability and Statistics," I review basic probability, including the concepts of den-sity functions, Bayes' theorem, maximum a posteriori (MAP) and ML estimators, change of variables and moments, and multivariate distributions. I also introduce a number of important distributions, including complex Gaussian, exponential, chi-square, and Rician distributions. I also discuss the concept of a random process. Finally, in Chapter 15, "Fourier Analysis," I review Fourier theory, including basic transforms, Fourier series, discrete-time Fourier transforms, and discrete Fourier transforms with a discussion of the usefulness of fast Fourier transforms (FFTs).

For an undergraduate semester class, I expect a typical presentation of the material would provide an overview of communications by using Chapter 2; quickly review the essential mathematic background by leveraging Chapters 13–15; and spend most of the class on Chapters 3–10. Depending upon the preparation of the class, I would include Chapter 11. Additionally, Chapter 12 might be presented for historical con-text. While some advanced concepts are introduced in Chapters 13–15, I provide these chapters primarily for review and reference, although the time dedicated to these chapters can be modulated depending upon the readiness of the students.

To support the classroom use of the textbook, I provide supplemental materials. These materials include a full set of slides for presentation, solutions for a subset of problems, and a rich set of supplemental simulation-focused problems. This material is available at www.cambridge.org/blisscomms.

Acknowledgments

It is my sincere hope that this text is useful, and I would like to thank all the students who made comments and suggestions. I would like to thank Rachel Lundwall, who provided valuable editorial comments. Particularly, I would like to thank my former student Professor Siddhartan Govindasamy, with whom I wrote my previous book. I learned much in that process. I would also like to thank all my friends at MIT Lincoln Laboratory and ASU, from whom I learned an immense amount over the years.

I would like to thank Cartel Coffee Lab, who helped fuel much of the writing of this text. The baristas are always friendly, and I still have not found a better espresso.

I would like to thank my parents, Daniel Bliss, Sr. and Nancy Bliss. I know that I got lucky because I still have not found better parents. I hope for my daughter's sake that I can be even a fraction of the parent that they were. Sure, I probably should have put them above the espresso, but, come on, espresso. I would also like to deeply thank Nadya and my daughter, Coco (who is fond of cats), for their love.

Part I

Communications Systems

1 Notation

In this chapter, we specify some of the notation that is used throughout the text. Many of these concepts will be described in greater detail in Chapter 13. To provide a common notational framework, we review basic mathematical concepts. We review tools for complex numbers, vectors, and matrices, and the relationship between exponentials and logarithms. We specify notation for integration. We discuss the relationship signal representation in terms of amplitude versus power and linear versus decibel.

1.1 Table of Symbols

In describing signals and parameters for communications, manipulation of scalars, vectors, and matrices are useful tools.

$$
\begin{array}{ll}
a \in \mathbb{S} & a \text{ is an element of the set } \mathbb{S} \\
a^* & \text{complex conjugate of } a \\
\mathbf{v}^T & \text{transpose of } \mathbf{v} \\
\mathbf{v}^H & \text{Hermitian conjugate of } \mathbf{v} \\
\mathbf{A}^T & \text{transpose of } \mathbf{A} \\
\mathbf{A}^H & \text{Hermitian conjugate of } \mathbf{A} \\
\forall x & \text{for all } x
\end{array}
\tag{1.1}
$$

We use $a \in \mathbb{S}$ to indicate that a is an element of a set. Common sets include the set of integers $\mathbb{Z}$, real numbers $\mathbb{R}$, and complex numbers $\mathbb{C}$. The complex conjugate of a number a^* flips the sign of the imaginary part of the number while leaving the real part unchanged. We often use vectors to represent sequences of numbers. The transpose of a column vector $\mathbf{v}^T$ converts the vector to a row vector, which we denote by $\underline{\mathbf{v}} = \mathbf{v}^T$. The Hermitian conjugate of a column vector $\mathbf{v}^H$ is given by the complex conjugate of all the elements of the transpose of the vector. Similarly, the transpose of a matrix swaps rows and columns $\mathbf{A}^T$ of the matrix, and the Hermitian conjugate of a matrix $\mathbf{A}^H$ is given by the complex conjugate of the elements of the transpose of the matrix. We review these ideas in greater detail over the next few sections.

1.2   Scalars

We indicate a scalar by a non-bold letter such as a or A. Scalars can be integer $\mathbb{Z}$, binary $\mathbb{Z}_2$, real $\mathbb{R}$, or complex numbers $\mathbb{C}$:

$$a \in \mathbb{Z},$$
$$a \in \mathbb{Z}_2,$$
$$a \in \mathbb{R}, \text{ or}$$
$$a \in \mathbb{C}, \tag{1.2}$$

respectively.

We indicate the square root of -1 by i,

$$\sqrt{-1} = i. \tag{1.3}$$

In some electrical engineering literature, this is denoted by j, but most fields use the notation i, so we conform to that broader tradition.

We can use the Euler number, e, to express values in the complex plane. The constant e is Euler's number that has an approximate value of $e \approx 2.718$. If we raise e to some real value x, then the result is the expected value, as we see in Figure 1.1. However, by using an imaginary exponent, we can specify points in the complex plane. Specifically, the Euler formula for an exponential of some real angle $\alpha \in \mathbb{R}$ in terms of radians is given by

$$e^{i\alpha} = \cos(\alpha) + i \sin(\alpha), \tag{1.4}$$

which we depict graphically in Figure 1.2.

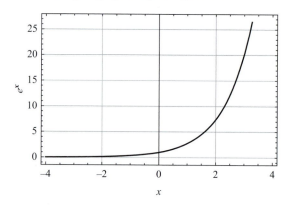

Figure 1.1 Graphical representation of the exponential.

Consequently, we can express a given complex number $a \in \mathbb{C}$ in terms of polar coordinates, as provided by the Euler formula, with a radius $\rho \in \mathbb{R}$ and an angle $\alpha \in \mathbb{R}$:

$$a = \rho \, e^{i\alpha}$$
$$= \rho \, \cos(\alpha) + i \rho \, \sin(\alpha). \tag{1.5}$$

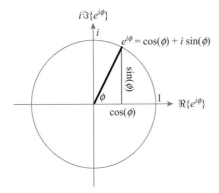

Figure 1.2 Graphical representation of the Euler formula in the complex plane.

We indicate the real and imaginary parts of the complex number a by

$$\Re\{a\} = \rho \, \cos(\alpha)$$
$$\Im\{a\} = \rho \, \sin(\alpha), \tag{1.6}$$

respectively. We indicate the complex conjugate of the variable a by

$$a^* = (\rho \, e^{i\alpha})^* = \rho \, e^{-i\alpha} \tag{1.7}$$
$$= (\rho \, [\cos(\alpha) + i \, \sin(\alpha)])^* = \rho \, [\cos(\alpha) - i \, \sin(\alpha)]. \tag{1.8}$$

The value of i can also be expressed in an exponential form:

$$i = e^{i\pi/2 + i 2\pi m} \; \forall \, m \in \mathbb{Z}. \tag{1.9}$$

We can use this form to evaluate raising i to some power. Because the value of the exponential is the same if you shift the phase by 2π, to be precise we include all possible additions of 2π. We often focus on the case of $m = 0$, but we should always be aware that other phases are possible. In considering the $m = 0$ case, the inverse of i is given by

$$\frac{1}{i} = i^{-1}$$
$$= e^{-i\pi/2}$$
$$= \cos(-\pi/2) + i \, \sin(-\pi/2)$$
$$= 0 - i. \tag{1.10}$$

We indicate the logarithm of variable $x \in \mathbb{R}$, assuming base $a \in \mathbb{R}$, by

$$\log_a(x), \tag{1.11}$$

such that

$$\log_a(a^y) = y, \tag{1.12}$$

under the assumption that a and y are real. When the base is not explicitly indicated, it is sometimes assumed that it indicates a natural logarithm (base e; Figure 1.3) such that

$$\log(x) = \log_e(x) = \ln(x). \tag{1.13}$$

In some texts, it is assumed that $\log(x)$ indicates the logarithm base 10 or 2 rather than the natural logarithm. For clarity, we explicitly include the base in this text $\log_e(x)$.

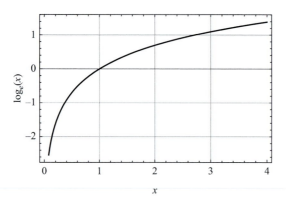

Figure 1.3 Graphical representation of the natural logarithm.

The translation between bases a and b of variable x is given by

$$\log_b(x) = \frac{\log_a(x)}{\log_a(b)}$$

$$= \log_b(a) \log_a(x). \tag{1.14}$$

We can expand the logarithm around 1 by using the argument $1 + x$ and expanding around $x = 0$ by using a Taylor series expansion about zero, which is given by

$$f(x) = f(0) x^0 + \frac{1}{1!} \left(\frac{\partial}{\partial x} f(x) \right) \bigg|_{x=0} x^1 + \frac{1}{2!} \left(\frac{\partial^2}{\partial x^2} f(x) \right) \bigg|_{x=0} x^2 + \ldots, \tag{1.15}$$

where we use $(\cdot)|_{x=0}$ to indicate evaluating the expression with $x = 0$. We can use this form to expand around $\log_e(1 + x)$ about $x = 0$, which is given by

$$\log_e(1 + x) = \sum_{m=1}^{\infty} (-1)^{m+1} \frac{x^m}{m} \;;\; \text{for } \|x\| < 1$$

$$\approx x \;;\; \text{for small } x. \tag{1.16}$$

Consequently, by using $x = y/n$ and Equation (1.16), we can show for a finite value of y:

$$e^y = \lim_{n \to \infty} \left(1 + \frac{y}{n} \right)^n. \tag{1.17}$$

For real base a, the logarithm of a complex number $z \in \mathbb{C}$, such that z can be represented in polar notation by using Equation (1.5),

$$z = \rho\, e^{i\alpha}$$

$$= \rho\, e^{i\alpha + i2\pi m} \;;\qquad m \in \mathbb{Z}, \tag{1.18}$$

is given by

$$\log_a(z) = \log_a(\rho) + \frac{i}{\log(a)} (\alpha + 2\pi m) ; \qquad m \in \mathbb{Z}. \tag{1.19}$$

For complex numbers, the logarithm is multivalued because any addition of a multiple of $2\pi i$ to the argument of the exponential provides an equal value for z. If the imaginary component produced by the natural (base e) logarithm is greater than $-\pi$ and less than or equal to π, then it is considered the principal value.

1.3   Vectors and Matrices

We indicate a column n-vector of complex values with

$$\mathbf{a} \in \mathbb{C}^{n \times 1}. \tag{1.20}$$

We indicate a row n-vector with a bold lowercase letter with an underscore, such as

$$\underline{\mathbf{a}} \in \mathbb{C}^{1 \times n}, \tag{1.21}$$

which has all n of its elements along one row. On occasion, we need to extract one element of a vector. The mth element in $\mathbf{a}$ is denoted

$$(\mathbf{a})_m \text{ or } \{\mathbf{a}\}_m. \tag{1.22}$$

We indicate a matrix with m rows and n columns by a bold uppercase letter:

$$\mathbf{M} \in \mathbb{C}^{m \times n}$$

$$= \begin{pmatrix} M_{1,1} & M_{1,2} & M_{1,3} & \cdots & M_{1,n} \\ M_{2,1} & M_{2,2} & M_{2,3} & \cdots & M_{2,n} \\ M_{3,1} & M_{3,2} & M_{3,3} & & \\ \vdots & & & \ddots & \\ M_{m,1} & M_{m,2} & M_{m,3} & \cdots & M_{m,n} \end{pmatrix}, \tag{1.23}$$

where $M_{p,q}$ is the element at the pth row and qth column of $\mathbf{M}$. If there is potential ambiguity, we find it useful to use the notation that operates on the matrix for the pth row and qth column of $\mathbf{M}$ that is given by

$$(\mathbf{M})_{p,q} \text{ or } \{\mathbf{M}\}_{p,q}. \tag{1.24}$$

1.4   Integrals

The area under some function is given by the integral over that function. To help keep track of variables for more complicated multiple integral scenarios, we use the convention that the differential is placed near the integration symbol. For example, the definite integral $\mathcal{I}$ of integrand $g(x)$ over the region bounded by a and b is written as

$$\mathcal{I} = \int_a^b dx\, g(x), \tag{1.25}$$

which we depict for some arbitrary function $g(x)$ in Figure 1.4. Furthermore, it is sometimes useful to think of integration as an operator that can be applied to the function $g(x)$; thus, this convention is useful there. However, the extent of the integrand may not always be clear, which is a disadvantage of this convention.

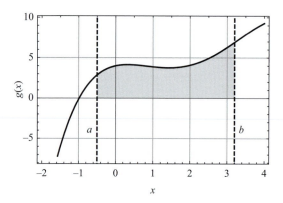

Figure 1.4 Notional depiction of integration given by the area under $g(x)$ from a to b.

A Dirac delta function (which technically is not a function) has zero width and infinite height with unit area and is generally used within the context of an integral [4]. It has the property of providing the value of the rest of the integrand at the point where the delta function is not zero. For example, with a real parameter a, the integral over the real variable x and well-behaved function $f(x)$,

$$\int_{-\infty}^{\infty} dx\, f(x)\, \delta(x - a) = f(a). \tag{1.26}$$

As a concrete example, we consider the integrand of $f(x) = 4x^3 - x^2 + ix + 2$ multiplied by $\delta(x - 1)$, so the integral is given by

$$\int_{-\infty}^{\infty} dx\, f(x)\, \delta(x - 1) = \int_{-\infty}^{\infty} dx\, (4x^3 - x^2 + ix + 2)\, \delta(x - 1)$$
$$= 4(1)^3 - (1)^2 + i\,(1) + 2$$
$$= 4 - 1 + i + 2 = 5 + i. \tag{1.27}$$

It is sometimes useful to identify situations in which two integers are the same. To do this indication, we use the notation for the Kronecker delta, which is given by

$$\delta_{m,n} = \begin{cases} 1 & ; \quad m = n \\ 0 & ; \quad \text{otherwise} \end{cases}. \tag{1.28}$$

1.5   Scales: Linear, Logarithmic, and Decibels

During the era in which calculations were performed with slide rules, the typical student of engineering was particularly good at converting between linear and
logarithmic scales. However, as we now use computers to perform most of our calculations, working with a logarithmic scale is slightly less natural. Nevertheless, it is
useful to practice. In communications, many system parameters are defined in units
of decibels. Examples include power, signal-to-noise ratio (SNR), attenuation, and
noise figure.

For some value p_{linear} expressed in terms of linear units, the value p_{dB} expressed
in terms of decibels is given by

$$p_{dB} = 10 \, \log_{10}(p_{linear}), \tag{1.29}$$

so the inverse relationship is given by

$$p_{linear} = 10^{p_{dB}/10}. \tag{1.30}$$

While there are counter-examples, values expressed on a decibel scale are often
related to power or manipulation of power. For an amplitude value x_{linear} that is
related to power by $p_{linear} = \|x\|_{linear}^2$, where $\|x\|$ indicates the absolute value of x,
the conversion between linear and decibel scale may convert to the power domain
implicitly, which is given by

$$\begin{aligned} p_{dB} &= 10 \, \log_{10}(\|x\|_{linear}^2) \\ &= 20 \, \log_{10}(\|x\|_{linear}). \end{aligned} \tag{1.31}$$

The conversion between linear and decibel scale is done so commonly that in most
communications discussion it is assumed that the reader will know to convert from
variables on a decibel scale to a linear scale before using them in an equation. The
student needs to be careful about this. Conversely, it is often convenient to implicitly
perform calculations on a logarithmic scale because multiplication and division are
mapped to addition and subtraction. Often mildly complicated evaluations of propagation loss and SNR can be done in one's head. Because of the value of moving
between linear and decibel scales, it is useful to have a number of conversions available in one's head. For example, moving between the values listed in Table 1.1 should
be automatic.

Notationally, we often specify a value on either linear or decibel scales and assume
that the reader will translate to the appropriate form as needed. For example, if we
specify values $x = 6$ dB and $y = 100$, then the product is

$$z = x \, y$$
$$= 4 \cdot 100 = 400 \approx 26 \text{ dB},$$
$$\text{or equivalently}$$
$$\approx 6 \text{ dB} + 20 \text{ dB} = 26 \text{ dB}. \tag{1.32}$$

Table 1.1 Linear and decibel scales.

Linear		Decibel
2	$\approx$	3 dB
$4 = 2 \cdot 2$	$\approx$	3 dB + 3 dB = 6 dB
10	$=$	10 dB
$100 = 10^2$	$=$	$2 \cdot 10$ dB
$1000 = 10^3$	$=$	30 dB
10^6	$=$	60 dB
$50 = 100/2$	$\approx$	20 dB $-$ 3 dB = 17 dB
$250 = 1000/4$	$\approx$	30 dB $-$ 6 dB = 24 dB

Problems

1.1 For some complex phasor $e^{i\alpha}$, evaluate the following:
(a) the real part of $e^{i\alpha}$
(b) the imaginary part of $e^{i\alpha}$
(c) $(e^{i\alpha})^*$
(d) $e^{i\alpha}$ for $\alpha = 0$
(e) $e^{i\alpha}$ for $\alpha = \pi$

1.2 Without explicit numerical evaluation, convert between logarithmic bases:
(a) Evaluate $\log_8(10)$ for $\log_2$.
(b) Evaluate $\log_2(e)$ for $\log_e$.
(c) Evaluate $\log_{17}(12)$ for $\log_{10}$.

1.3 Show that the following is true for finite x:

$$e^x = \lim_{n \to \infty} \left(1 + \frac{x}{n}\right)^n.$$

1.4 Convert the following linear power values to an approximate decibel value:
(a) 10
(b) 2000
(c) 400
(d) 50
(e) 100^3
(f) $(20)^3$

1.5 By using basic knowledge of linear and dB conversions for power signals, express the following items from linear to dB or dB to linear as needed. Show your work.
(a) 4000
(b) -17 dB
(c) $\sqrt{5}$
(d) 93 dB
(e) $\frac{1}{\sqrt{20}}$

1.6 Evaluate the integral

$$\int_{-\infty}^{\infty} dx \, \frac{\delta(x-3) \cos^2(4\pi x)}{1 + \sin(2\pi x)}.$$

1.7 Evaluate the integral

$$\int_{-\infty}^{\infty} dx \, (1 + x^2 + x^3) \, \delta(2x - 1).$$

2 Basic Radio

In this chapter, we discuss the essential ingredients of a wireless communications system. While we focus on these radio systems, the underlying concepts apply to a wide range of communications systems, including those that communicate via acoustics, wires, and fiber optics. We introduce the idea of a communications link that uses the electromagnetic spectrum. We discuss sources of data that we communicate. We introduce the idea of the theoretical limit on data rates, given the communications channel, and provide an introduction to the characteristics of a simple channel, including propagation loss and noise. We provide an overview of the basic components of a radio system for the transmitter and receiver. We review the meaning of the Nyquist rate, and how we group signal energy into the overlapping ideas of samples, chips, symbols, or channel usage.

2.1   What Is Communications?

It is so common to modern life that we sometimes take it for granted, but what is meant by *communications* has changed dramatically over the last century. Whenever we have a concept that we want to transfer to someone else, we have a communications problem. At some point in the distant past, evolution gave our ancestors verbal language. At that point, we could communicate complicated ideas to each other. This must have been a remarkable transition and a huge evolutionary advantage. Nearly 5000 years ago, both the Egyptians and Sumerians developed written language. Concepts could be transferred relatively directly even if the individuals were separated by time and space. For a long time, writing your thoughts down and mailing them was your best bet. In 1837, William Cooke and Charles Wheatstone – and, separately, Samuel Morse – filed patents for telegraph systems, which allowed for quick transmission of alphanumeric messages over long distances over wire. In 1876, both Alexander Bell and Elisha Gray developed the fundamentals of telephone technology to communicate analog voice signals over wires and filed for patents essentially at the same time.

A number of engineers and scientists made progress on wireless communications technology in the years around 1900, and Guglielmo Marconi developed it into a viable commercial enterprise. This and several hundred other important technological developments inevitably led us to the pinnacle of human achievement:

the ability to watch cat videos on our phones no matter where we are (for the most part).

2.1.1   A Communications Link

We have a source of information (which we often describe in terms of some sequence of bits) at one location and a destination (or sink) at another. We use a transmitter to send a signal containing the information and a receiver to capture this signal and extract the information, as seen in Figure 2.1. The thing in between these two entities is the channel. For any physical system, this channel distorts whatever signal (which contains the information) we try to send through it. The channel may be a wire or, as we will consider primarily here, electromagnetic waves propagating through the environment. We commonly identify the range of electromagnetic wave frequencies from just above around 3 kHz to less than 300 GHz as radio frequencies (RF).

Figure 2.1 Notional wireless communications link that is sending a signal from the transmitter to the receiver.

For wireless systems, we often construct a signal at complex baseband, which is spectrally centered around zero frequency or, equivalently, around DC,[1] and up-convert the signal to passband, which is centered at the carrier frequency. There is also a component around the negative carrier frequency. The carrier frequency is typically large compared to the bandwidth of the baseband signal. At the receiver, the signal is down-converted back to a complex baseband signal, which is spectrally centered around DC again. While the signal that goes over the air at the carrier frequency is real, for idealized transmit and receive processing chains (discussed in greater detail in Section 2.5), all the information of the passband signal is represented at complex baseband. Consequently, unless we are specifically considering some particular distortion associated with the electronics of the radio system, we only need to consider the complex baseband signals at the transmitter and the receiver. The up- and down-conversion employs mathematical tools discussed, in part, in Section 15.4.2. This concept is described in greater detail in Chapter 5.

2.1.2   Signal Sources and Coding

While we will revisit source signal construction and encoding later (in Section 3.4.1) in more technical detail, it is worth providing a short introduction. Some signals

---

[1] DC (direct current) is used to indicate a constant signal or, in other terms, the amplitude associated with zero frequency.

are intrinsically digital. For example, if I need to send an email, text message, or a sequence of entries in financial accounts, the data can be translated directly to a sequence of binary ones and zeros based upon a given deterministic mapping. At the receiver, we can map these ones and zeros back to the original text message, or whatever the digital message was.

If the signal is naturally analog, such as spoken language or a visual image, it must be converted to a digital signal to use modern digital communications. Each analog signal type has a different digitization chain, but typically there is some analog processing (such as amplification and filtering) that is followed by an analog-to-digital converter (ADC). In the case of an image, analog processing is performed by lenses that project a focused image on a sensor. The lenses are followed typically by an array of color filters covering an array of light-sensitive detectors. The intensity of light is measured and converted to a set of discrete levels for each cell on the array (which effectively digitizes the image).

Given a signal source, we observe that the number of bits used to encode the signal is often wasteful. Thus, it is in the communications system's interest to perform source encoding to reduce or compress the number of bits. This coding or compression can broadly fall into two classes: lossless and lossy. Lossless source coding is exactly reversible. That is, there is no information loss in performing the compression. Conversely, lossy compression is willing to lose some information to significantly reduce the number of bits. In both cases, knowledge of the source probability density function is required to perform this compression well.

As an example, for cellular phones lossy voice compression is performed by the vocoder (voice source compression coder), which employs a prior knowledge of human voice and language signal distributions to compress the signal dramatically with only mild distortion. For example, audio sampling is often performed at a rate of greater than 40,000 samples per second. Each sample may have 16 bits or more. Consequently, we have an uncompressed source data rate of at least 640,000 b/s. It is common for vocoders to compress this rate down to about 10 kb/s with only mildly noticeable distortion. When compressed down to 5 kb/s, the distortion is typically noticeable.

2.2   Radio Performance Limit

As we discuss in greater detail in Chapter 3, if we correct for all other distortions in a system, we are left with the thermal noise at the receiver. The fundamental limit of communications performance is given by the Shannon limit. It is a simple function of the signal-to-noise ratio (SNR) at the receiver. This achievable limit on data rate is one of the most remarkable and elegant results in all of engineering, so we preview it here:

$$C \leq B \log_2(1 + \text{SNR}), \tag{2.1}$$

where C is the bounding data rate (b/s) and B is the full bandwidth.

2.2.1   Radio Problem: Distortion of Received Signal

There are a variety of sources for this distortion to our transmitted signal. Any real transmitter and receiver is always nonlinear; consequently, amplitude fluctuations of the signal are always distorted to some extent. There is always some, although potentially a very small, amount of dispersion. Dispersion is caused by memory or delay spread in the channel. In the frequency domain, this dispersion means that the channel is frequency-selective, that is, the amplitude and phase of the attenuation vary as a function of frequency. The effect of this dispersion is that the current received symbol is received with echoes of previous symbols superimposed on it. This effect is known as inter-symbol interference (ISI). There also may be external interference caused by RF sources unrelated to the transmitter of interest that impinges upon the receive antenna simultaneously at the same frequency as the signal of interest. Finally, and most fundamentally, there is thermal noise at the receiver.

2.2.2   Propagation Attenuation

The most fundamental parameter that determines how well a communications system can perform is the SNR at the receiver. The SNR is the ratio of the signal power P_{signal} to the noise power P_{noise}, given by

$$\text{SNR} = \frac{P_{signal}}{P_{noise}}. \tag{2.2}$$

At every point in the receiver processing chain, there is an effective SNR, so there can be some confusion about which SNR is being referenced. For the discussion in this section, we consider the SNR at the input of the first amplifier, which we often denote a low-noise amplifier (LNA). For a given bandwidth, more data can be sent as the SNR increases. The channel between the source or transmitter and the sink or receiver reduces whatever signal power was present at the transmitter. Simply put, the vast majority of photons leaving the transmitter do not make it to the receiver. This attenuation affects the signal power observed at the receiver. Every channel has different propagation characteristics, so there is no one answer. If we parameterize the complex amplitude with parameter a, then the attenuation in power is given by $\|a\|^2$.

For a simple flat-fading channel (which is one without observable spread in delay), the basic model is given by

$$z(t) = a\,s(t) + n(t), \tag{2.3}$$

where $z(t)$ is the received signal, $s(t)$ is the transmitted signal, and $n(t)$ is the additive Gaussian noise at the receiver for which we have ignored the effects of the propagation delay. Usually we assume that the received signal is filtered to match the bandwidth of the signal of interest to minimize the out-of-frequency-band noise power. At this point, it is worth mentioning a fundamental ambiguity. In certain situations, such as understanding the physics of the channel, we consider a to include only the effects of the physical channel between antennas. In this case, the variance of $s(t)$ given by $E[\|s(t)\|^2] = P_t$ is the power at the transmit antenna. In other scenarios, such as analysis of receivers, it is useful to include the transmit gain as part of the channel so that $\|a\|^2$ is the gain-attenuation product, under the assumption $E[\|s(t)\|^2] = 1$. We use both conventions. The situation determines which convention we use, and we expect that the usage will be clear from the context. For the remainder of this chapter and for Chapter 3, we use the convention that $\|a\|^2$ is associated with the propagation attenuation.

The power attenuation for a basic wireless line-of-sight channel is given by the Friis equation,

$$P_r = \|a\|^2 P_t \tag{2.4}$$

$$\|a\|^2 = \frac{G_t G_r}{(4\pi r/\lambda)^2}, \tag{2.5}$$

where G_t is the antenna gain of the transmit antenna in the direction of the receiver, and G_r is the receive antenna gain in the direction of the transmitter. We consider non-line-of-sight models in Section 7.6.1. The wavelength is indicated by λ and the distance from the transmitter to the receiver is indicated by r. For electromagnetic waves, these gains include the effects of polarization of the antennas.

For terrestrial wireless communications, there are often scatterers in the environment that affect the propagation. While there are counter-examples, the effect of these scatterers typically reduces the signal power at the receiver compared to line-of-sight propagation; consequently, $\|a\|^2$ is smaller than that indicated by the Friis equation (Equation 2.5). There are many models that attempt to incorporate this effect, but a simple one is power law attenuation given by

$$\|a\|^2 = k \frac{1}{(r/\lambda)^\alpha}, \tag{2.6}$$

where k is some constant of proportionality, and α is the power law exponent. For line-of-sight, α is 2. For different environments, one can vary α as an approximation of the types of attenuation that are expected. In urban environments, a value of $\alpha \sim 4$ is often assumed. Although this approximation has been supported by various experimental results, it is always a rough approximation.

2.2.3   Antennas

The antenna gain expressed in Equation (2.5) is an expression of the directionality of the radiated electromagnetic energy. An antenna converts the time-varying electrical signal of the currents and voltages in an electrical cable to radiated electromagnetic waves. The way in which antennas convert signals to radiated waves is complicated, but if we consider the propagating waves from the antenna we view the signal as radiating with a given polarization and intensity as a function of direction. Importantly, antennas have some range of useful frequency. Outside of this range, the antennas are very inefficient. Consequently, antennas have to be carefully chosen to match the problem and provide some filtering.

An easy antenna to describe is the dipole [5]. As seen in Figure 2.2, an oscillating voltage drives a signal into two wires that lie along a line. Because the wire is open, there is nowhere for the current to flow, but the infinitesimal motion of the carriers is sufficient to radiate electromagnetic power. As the voltage swings back and forth, a signal is electromagnetically radiated with azimuthal symmetry about the line in which the antenna lies. The maximum power is radiated uniformly azimuthly about the line of the antenna. No power is radiated in the direction along the antenna. If the antenna is oriented vertically, then we say that the radiated waves are vertically polarized. If the antenna is oriented horizontally, then we say that the radiated waves are horizontally polarized. Circular or elliptical polarizations can be constructed by using both vertical and horizontal polarizations and sending signals out of phase. Other polarizations can be constructed by using more complicated geometries.

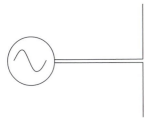

Figure 2.2 Simple example of a voltage driving a dipole antenna.

We often compare antennas to a notional, physically unrealizable isotropic antenna. This notional isotropic antenna radiates the same power in all directions. This is not possible because some directions are always preferred by the antenna. As an example, the dipole antenna prefers to radiate along a direction perpendicular to the antenna, as seen in Figure 2.3 for a short dipole that radiates power $P_{rad}(\phi, \theta)$ as

$$P_{rad}(\phi, \theta) = \cos^2(\theta), \tag{2.7}$$

where ϕ is the azimuthal angle measured around the line of the antenna, and θ is measured from the direction perpendicular to the antenna. Note that the radiated power is not a function of ϕ. If we average the radiated power over all directions from a given antenna, then we can compare it to an isotropic antenna. If we feed unit power into the isotropic power, then a given power density is radiated in any direction. For a given direction and polarization, the power density of a real antenna will have some value. We say that the ratio of the real power density to that of the isotropic antenna is the gain for that direction and polarization. Often, as a short hand, we will say that the gain of an antenna is the direction and polarization with the maximum gain. However, the gain that is important for Equation (2.5) is the gain from one antenna to the other. The gain (peak gain) of a dipole antenna is about 1.76 dBi. The unit dBi means dB relative to an isotropic antenna. Interestingly and conveniently, because of reciprocity, the receive gain is the same as the transmit gain. Consequently, you only have to know one number.

There is a relationship between the size of antenna and the largest possible gain. If one thinks of a receive antenna as something that catches photons, the number of

Figure 2.3 Antenna pattern of a short dipole. The radius of the shape indicates the antenna gain in that direction. The dipole antenna is included for geometric reference.

photons that one could catch grows with the area of the antenna perpendicular to the direction of propagation. This heuristic explanation clearly falls apart when we think about a dipole that has effectively no area, but it does sort of work for larger antennas. The peak gain G for an antenna is related to the effective area A_{eff} by

$$G = 4\pi \, \frac{A_{\text{eff}}}{\lambda^2}, \tag{2.8}$$

where λ is the wavelength. For well-designed higher-gain antennas, the effective area is close to, if slightly smaller than, the physical area.

As an example, consider the parabolic dish we depict in Figure 2.4. A parabolic dish uses a small antenna feed at the focus of the parabola. The dish is part of the parabola that reflects lines radiating from the feed to parallel lines radiating from the dish. Because of the shape of the parabola, the total path length of each ray, from a point in the distance to the reflecting dish to the focus are equal. The edges of the dish do not work properly because induced currents are disturbed by the discontinuity. Because of reciprocity, reception works the same way, except with time reversed. The feed is relatively small, but the gain is mostly set by the size of the reflector. Because the antenna is based upon simple geometry, these antennas can operate over a wide range of frequencies. The National Radio Astronomy Observatory's Very Large Array uses 28 25-meter diameter parabolic antennas. The gain of each antenna is set by the 25 m dish diameter. The system can operate at center frequencies from 1 to 50 GHz. For example, let's assume that the center frequency is $f_0 = 10$ GHz. If we assume that the effective area of one dish is the physical area, the gain is given by

$$
\begin{aligned}
G &= 4\pi \, \frac{A_{\text{eff}}}{(c/f_0)^2} \\
&= 4\pi \, \frac{\pi \, (25 \text{ m}/2)^2}{(c/[10 \cdot 10^9 \text{ Hz}])^2} \\
&\approx 6.9 \cdot 10^6 \approx 68 \text{ dB},
\end{aligned}
\tag{2.9}
$$

where c is the speed of light. This estimate of gain is a little optimistic in that the actual peak gain is lower, but this calculation provides a way to estimate approximately the gain from the geometry.

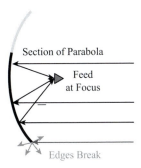

Figure 2.4 Notional example of how a parabolic dish works.

2.2.4 Thermal Noise

The fundamental culprit for distorting our signal at the receiver is additive noise, which is caused by the thermally induced motion of charged particles, as seen in Figure 2.5. The fundamental problem is how much information we can send, given these channel distortions and the limits on our resources. As there are a huge number of randomly moving charged particles in any macroscopic sample of material, it is not surprising that the observed noise signal is drawn from a Gaussian distribution. This intuition comes from the application of the central limit theorem. If we consider the thermal noise in our receiver a complex baseband signal, the noise power spectral density as a function of frequency f is given by [6]

$$\rho_{noise}(f) = \frac{hf}{e^{\frac{hf}{k_B T}} - 1} \tag{2.10}$$

$$\approx \frac{hf}{1 + \frac{hf}{k_B T} - 1} = k_B T; \quad \frac{hf}{k_B T} \ll 1, \tag{2.11}$$

where $h \approx 6.626 \cdot 10^{-34}$ J s is the Planck constant, $k_B \approx 1.38 \cdot 10^{-23}$ J/K is the Boltzmann constant, and T is the absolute temperature, expressed in units of Kelvin here. This effect is often denoted Johnson–Nyquist noise. For most current engineering scenarios, the approximation $hf/(k_B T) \ll 1$ is very good. The region of validity of the approximation can be observed in Figure 2.6. Below a few terahertz of frequency, the approximation of a spectrally flat response is an excellent approximation. Under this approximation, the noise power P_{noise} for some total bandwidth B is given by

$$P_{noise} = k_B T B, \tag{2.12}$$

where the bandwidth B indicates the total bandwidth around some carrier at passband, or equivalently, the spectral width of the signal including both positive and negative frequencies at complex baseband.

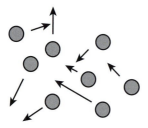

Figure 2.5 The thermal excitation of charged particles introduces noise.

At complex baseband, we have complex Gaussian noise. A given noise sample n at some time is drawn from the zero-mean circularly symmetric complex Gaussian distribution (described in Section 14.1.6), which is described by the probability density function $p(n)$

$$p(n)\, dn = \frac{1}{\pi\, \sigma^2}\, e^{-\|n\|^2/\sigma^2}, \tag{2.13}$$

where the variance of the noise is just the noise power, which is given by

$$\sigma^2 = P_{\text{noise}} = k_B\, T\, B. \tag{2.14}$$

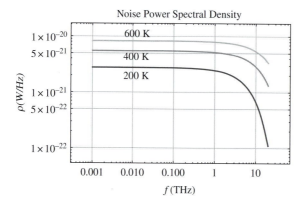

Figure 2.6 The power spectral density of thermally induced noise as a function of frequencies for receiver temperatures of 200 K, 400 K, and 600 K.

2.2.5 Filtering

Implicit in the discussion of noise power (and thus of SNR, which was discussed in Section 2.2) was an assumption that the noise was filtered down to some bandwidth.

There are analog and digital filtering approaches, and both are typically employed by communications systems. In both cases, the spectrum of the signal being filtered is reshaped. The shaping can be arbitrary, but most employed filters are in one of three categories – low-pass, high-pass, or band-pass – which indicate what range of frequencies are allowed to pass through the filter. In the case of the complex noise represented at complex baseband, a low-pass filter is employed with a corner frequency of $B/2$, so that the total complex baseband bandwidth, including both positive and negative frequencies, is B. Ideally, the portions of the signal outside this band at frequency $\|f\| > B/2$ are reduced to essentially zero, while the portion of the signal with frequencies within the band $\|f\| \leq B/2$ is allowed to pass without modification. In real applications, this transition cannot be so abrupt.

Conceptually, digital and analog filtering are related. Digital filtering combines delayed and scaled copies of the signal to produce the desired spectral effect. We discuss these filters in greater detail in Section 15.8. For analog circuits, there are multiple approaches. One can construct filters from the appropriate connection of inductors, capacitors, and resistors. Alternatively, electromechanical devices can be employed, such as surface acoustic wave (SAW) filters that exploit mechanical wave propagation via piezoelectric couplings to a crystal or ceramic. The filtering is produced by combining multiple paths.

2.3   Radio Control

2.3.1   States of Radio Operation

A radio cycles through various states of operation. We can express the transitions between these states as a finite-state machine. An example of a simple radio finite-state machine is displayed in Figure 2.7. The power-up operation is probably self-explanatory, but in practice there can be a complicated set of system calibrations at startup. The radio may simply wait and try to acquire a beacon (a transmitted signal used to notify other nearby radios of its existence), or it may transmit a beacon itself. Once two radios have established each others' existence, timing and frequency synchronization occurs (although this may be done at acquisition). The radios may then transmit and receive packets of information. Communications is often not continuous, so radios often stop transmissions or go into a mode in which just link maintenance and control data are communicated. Synchronization information is often tracked during this state. Finally, the radio may power-down.

2.3.2   OSI Model

In Figure 2.8, the layers of the open systems interconnection (OSI) model (or informally denoted the OSI stack) are depicted. While many systems do not fully populate these layers or blur the lines between them, the framework is a useful way to formalize all the pieces needed for a typical communications system. Furthermore, if the

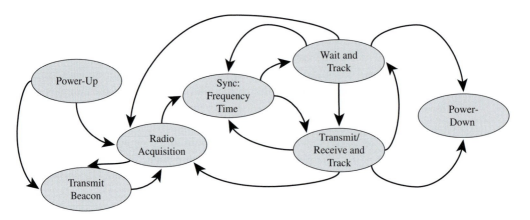

Figure 2.7 Basic finite-state machine for radio operation.

structure of the model is observed, then it is relatively easy to modify details within a layer without affecting other layers.

The application layer at the top indicates how the information is used. This text-book focuses on the lower layers – primarily on the physical layer but with some discussion of the data link layer. The data link layer focuses on the control of the link between the transmitter and the receiver. It controls the resending of frames of data, if required. It also includes the media access control (MAC). The physical layer trans-lates how the underlying bits are mapped into symbols and the eventual over-the-air signal. This layer includes forward error correcting codes.

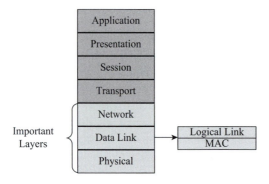

Figure 2.8 Layers of the OSI model.

2.4   Waveform Frames

In a typical communications signal, the form of the information being sent must be defined. In general, there are three classes of information being sent, as indicated in Figure 2.9. One class is the data of interest, as you would expect. The second class is that associated with training data or, equivalently, pilot signals or pilot sequences. This information is known at both the transmitter and the receiver and is used for channel estimation and synchronization. The last class is associated with control signals. For many communications systems, a significant set of control signals is required to enable the communication system to operate.

What's in Your Frame or Packet

Control Data	Pilot	Your Data

Figure 2.9 Contributions to the transmitted signal organized as a frame or packet.

2.5 Transmit and Receive Processing Chains

A radio (or, equivalently, wireless communications) system must address a number of issues. In this discussion, we only consider the lower layers of the OSI model (described in Section 2.3.2). A radio transmitter (whose notional transmit processing chain is depicted in Figure 2.10) must have some information to send. Hopefully, this data has been compressed so that bits are not sent wastefully. The radio system has to decide when and at what carrier frequency it can send the signal. It must convert the information bits into a signal that can be transmitted from the antenna. There are a number of issues that need to be addressed in making these decisions.

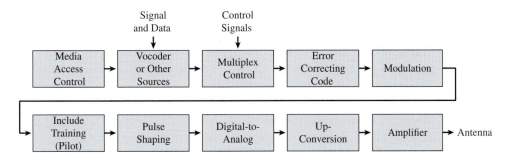

Figure 2.10 Notional chain of functional blocks for a radio transmitter.

At the receiver, these steps need to be reversed so that the underlying information bits can be decoded. A typical receive chain is depicted in Figure 2.11, which corresponds to the transmitter in Figure 2.10. Not all radios have all these components, and some radios have more components, but these chains are representative.

2.5.1 Receive Processing Blocks

In explaining the overall structure of transmit and receive processing chains, it is unclear whether we should start with the transmitter or the receiver. The reality is

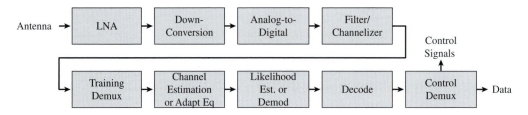

Figure 2.11 Notional chain of functional blocks for a radio receiver.

that their functions are inextricably tied. For the discussion within the main portion of the text, we will describe them in paired blocks; however, here we will start with the receiver chain.

As seen in Figure 2.11, the signal is received by an antenna. Antennas convert electrical signals to and from electromagnetic waves. If you consider it for a moment, it seems quite magical. The impinging electromagnetic waves induce currents in the antenna, which are then observed as an electrical signal across a couple of physical points on the antenna.

As seen in Figure 2.11 and discussed generally in Chapter 4 and specifically in Section 4.3, the first block in the receiver is typically an LNA. The LNA is used to minimize the effects of thermal noise at the receiver. Noise within the receiver is dominated by the first device in the processing chain. By having a reasonable LNA first, the effect of noise introduced by subsequent devices is reduced.

Because it is desirable to perform processing at complex baseband rather than passband, the next stage in the processing chain is typically a down-converter that transforms the signal near the carrier frequency to frequencies around DC, as seen in Figure 2.11 and discussed more generally in Chapter 5. There are multiple approaches to perform this operation. In modern radios, it is common to use direct conversion.

While the down-conversion may be performed partly or completely in the digital domain, it is common to convert the analog signal to the digital domain at complex baseband (Figure 2.11). Consequently, in this case, there are two ADCs used. One is associated with the real component and the other with the imaginary component of the resulting complex baseband signal. This conversion is discussed in greater detail in Section 5.7.

In the next step of the receive processing chain, some sort of filtering or chan-nelizing is typically performed (Figure 2.11). There are a variety of reasons for this operation. It may be simply employed to filter out-of-band noise, which is often performed along with some resampling to map the received sample rate to a new desired sample rate. It may compensate for the pulse-shaping filter used by the transmitter. If the modulation employs a multiple-carrier approach, such as orthogonal frequency-division multiplexing (OFDM), then this stage may include channelizing. For OFDM, this channelization is performed by a fast Fourier transform (FFT).

Typically, there is some training data embedded within the transmitted signal. The next step in the processing chain is to extract this portion of the data, as seen in Figure 2.11 and described in Section 6.3.1 and in Chapter 7. This data can be used for transmitter acquisition and for timing and frequency synchronization. Its primary role is to be used as training data to estimate the effects of the channel. This channel estimate is used by the following stage.

The next step in the processing chain is demodulation, which converts baseband signal voltages to symbols, or at least likelihoods of symbols (as seen in Figure 2.11 and described in Section 6.3 and Chapter 7). At complex baseband, various symbols were sent. We now have noisy observations of these symbols. As system designers, we have two basic choices at this point. We can make hard decisions, so that estimates of which symbol were sent are determined. Alternatively, as is often done in modern communications systems, we can evaluate estimates of likelihoods for the set of possible symbols. In either case, these estimates are passed to the decoder of the forward error correcting code.

The next step in the processing chain is to extract estimates of the original information and control bits (as seen in Figure 2.11 and described in Chapter 8). In modern communications systems, parity information, which carries redundant representations of the original information bits, is incorporated into the transmitted signal by forward error correcting codes. By exploiting this parity information, the decoder can significantly improve the fidelity of the estimate of the transmitted information bits.

2.5.2 Transmit Processing Blocks

The transmit chain approximately reverses the order of the receive processing chain. At the transmitter, as observed in Figure 2.10, the first block we consider is the MAC. We discuss this block in greater detail in Chapter 10. This controls the details of how the radio should transmit to avoid adverse effects on other communication links.

The next block provides the source data that we wish to transmit (Figure 2.10). This data may come from vocoder or other sources of data. Typically, some form of source compression is performed to reduce the number of bits that need to be sent. We discuss source compression in greater detail in Sections 2.1.2 and 3.4.1.

The next block in Figure 2.10 multiplexes a control signal into the data stream. In real radio systems, to maintain the communications link, a number of control signals are required that have little to do with the information that needs to be sent. Some examples of these signals include radio status, requests for future channel usage, power control information, and frequency correction parameters.

The next block in Figure 2.10 is the application of forward error correction. In this block, the radio applies error correcting codes to compensate for channel distortion and noise. These codes increase the number of bits by adding parity bits. Strong codes can approach the performance predicted by the Shannon limit. We discuss error correcting codes in greater detail in Chapter 8.

At complex baseband, the modulation block in Figure 2.10 translates the bits to a set of complex amplitudes or symbols. This set is often called a constellation and is typically discrete, at least under the correct transformation. As an example, OFDM transmits a sequence in time that is constructed by applying an inverse FFT to a sequence of modulated symbols. Consequently, an FFT must be applied at the receiver to recover the modulated symbols. We discuss modulation in greater detail in Chapters 6 and 7.

Most radio systems employ some training or pilot sequence to help estimate the channel between the transmitter and the receiver. The training block in Figure 2.10 multiplexes symbols known to both the transmitter and the receiver into the sequence of symbols being transmitted. At the receiver, these symbols are employed to estimate channels, timing, and frequency offsets. Compensating for timing and frequency offsets is often denoted *synchronization*. We discuss the use of training data in greater detail in Section 6.3.1 and in Chapter 7.

While pulse shaping is not technically required, many communications employ it (Figure 2.10) to reduce the spectral spread of the transmitted signal. If one considers a single-carrier modulation, then each symbol would naively occupy a duration of $T \approx 1/B$; however, simple modulation approaches cause much larger spectral use. This spectral spread can be mitigated by convolving the symbol with a pulse-shaping filter. This concept is discussed in greater detail in Section 6.2.1.

The next step in the transmit processing chain represented in Figure 2.10 is the conversion of the signal from a digital to an analog representation. If the signal is expressed at complex baseband, then the real and imaginary terms (also designated *in-phase* and *quadrature*, respectively) of the signal require two digital-to-analog converters (DACs). This concept is discussed in greater detail in Chapter 5.

If there is an analog up-conversion stage, which is typical and is depicted in Figure 2.10, then the radio moves the complex baseband or intermediate frequency (IF) signal to the passband. This operation is performed by multiplying (or mixing) the input signal with the appropriate frequency to move the spectral location of the signal to the passband. This process may be followed by filtering to remove unwanted spectral contributions. Up-conversion is discussed in greater detail in Chapter 5.

Within the digital processing components of the chain, the signal is typically represented as complex baseband; however, this is not required because systems sometimes up-convert the signal digitally and send either an IF or a passband version of the signal to the DAC if the DAC is fast enough. By doing this digital up-conversion, only a single DAC is required. For the poorly named HF (high-frequency)[2] communications, having the DAC produce the passband signal is quite viable.

Finally, the passband signal is sent to an amplifier and then to an antenna (Figure 2.10). The power amplifier is discussed in more detail in Section 4.2. While it is often omitted in an introduction to communications, it is worth noting that both the

---

[2] It is poorly named because the so-called HF band (3–20 MHz) is now considered relatively low frequency.

amplifier and antenna can have significant effects on the passband signal. In some cases, this distortion can cause issues for the receiver. The amplifier is theoretically a perfectly linear device that does not affect signal in any way except by increasing its amplitude. In practice, all amplifiers are nonlinear and have some memory (or filtering effect). The nonlinearity is often represented by a Taylor expansion of the functional relationship between the input and output signals. Unsurprisingly, as the output signal approaches the levels of the amplifier's supply, it becomes increasingly difficult to maintain amplifier linearity. Furthermore, there is often a trade-off between the significance of the nonlinearity and the amplifier's power consumption. Despite manufacturers' efforts, there are always some memory or filtering effects. These effects can be a function of the signal voltage, which mixes the nonlinear effects and the filtering effects. These combined effects can be represented by a Volterra series expansion.

As we said previously, antennas seem magical. They convert information, represented as an electrical signal, into photons flying through space. The alternating voltages applied across these devices cause tiny motions of carriers of electric charge. These alternating currents induce electromagnetic radiation, which streams away from the antenna. The detailed structure of the antenna determines in what directions and polarizations this radiation preferentially flows as a function of frequency, that is, the antenna has some radiation pattern and frequency response (filtering).

2.5.3 Nyquist Rate

If we have a band-limited signal and know its values at some required regular sample rate, then we can exactly reconstruct this temporal signal. The minimum regular sample rate for this to be true is the Nyquist rate. As the Nyquist rate is typically introduced, we are often told that this rate is twice the single-sided bandwidth. However, the use of the word *bandwidth* is potentially ambiguous. We define the signal bandwidth to be the double-sided or full bandwidth (see Figure 2.12). Consequently, the Nyquist sample rate R_{nyq} is given by

$$R_{nyq} = 2 B_{ss}$$
$$= B, \tag{2.15}$$

where B_{ss} is the single-sided bandwidth and B is the double-sided bandwidth. When we shift the baseband signal up to passband for transmission, we directly observe double-sided or full bandwidth B, so it is more common to use this definition in communications.

2.5.4 Rate of Information

We often describe information rates in terms of bits per second (b/s). Less commonly used is nats per second. One nat is given by the inverse of the natural log of 2, so 1 nat $= 1/\log_e(2)$ b ≈ 1.44 b. The unit of duration for the transfer of information is given by a second in these definitions. We could have used nanosecond or century,

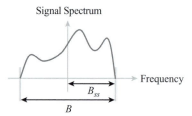

Signal Spectrum

Frequency

B_{ss}

B

Figure 2.12 Comparison of bandwidth definitions for band-limited signals.

but second is pretty natural. Each of these can be converted to seconds if we know enough about the system.

We may wish to describe rates of data in terms of alternative units: bits per sample, bits per symbol, bits per second per hertz, bits per chip, or bits per channel usage. These definitions overlap depending upon the system.

Sample: An essential observation is that, in modern systems, we perform much of the processing digitally. Consequently, for both the transmitter and the receiver, we operate on digital representations of the signal that is sampled in time. For most systems, these samples occur at regular intervals at some sample rate. Because we perform most of our processing at digital baseband, these samples are typically complex. We can imagine that representing the number of bits per sample would be a useful metric. However, there is some potential ambiguity. While we will not discuss this in detail within this text, we often operate in an oversampled regime. We always apply some filtering to our signals, and Nyquist tells us that if we have a band-limited signal, then sampling faster than the Nyquist rate provides no more information. However, for practical reasons, it is often useful to sample faster than Nyquist would suggest by some factor.

Chip: Even if we oversample so that our sample rate is higher than that required by Nyquist, we can imagine resampling (at least conceptually) the oversampled signal so that our new samples are at the Nyquist rate. For the double-sided (or equivalently full) bandwidth, these complex samples arrive at a rate equal to B. We often refer to this rate as the chip rate, and to the time between these samples as a chip. As we usually formulate the communications system discussion, the thermal noise is independent from chip to chip. This nomenclature is particularly common when using code-division multiple access (CDMA) approaches (discussed further in Section 10.3.4), but is more broadly useful.

Symbol: The use of the word *symbol* is particularly flexible: what is meant depends upon the topic we are discussing. As we begin our discussions, we will often assume that a symbol is sent for each chip. However, for both OFDM and CDMA, symbols are spread across many chips. We hope that we can determine what is meant by *symbol* from the context, but it is worth keeping in mind its varied usage.

Channel usage: Information theorists often discuss information transfer in terms of channel usage. By using this term, we can minimize what we need to know about the system. A channel usage could be a letter written on a dirty piece of paper or an attempt to send information given one chip. However, if information is spread across many samples, a channel usage could be the sequence of samples. The same applies if the information is spread across multiple samples in frequency. The concept is pretty flexible.

Problems

2.1 Represent the following in terms of dBW and dBm:
(a) 1 W
(b) 1 mW
(c) 5 W
(d) 100 mW
(e) 4 μW
(f) 20 kW

2.2 Evaluate the thermal noise power in watts and dBW for the following:
(a) $B = 10$ MHz, $T = 300$ K
(b) $B = 1$ GHz, $T = 400$ K
(c) $B = 1$ MHz, $T = 77$ K

2.3 For an audio signal digitized at 44,000 samples per second with 16 bits per sample,
(a) evaluate the required raw bit rate, and
(b) calculate the compression ratio (that is, the ratio of actual to raw rates) if a vocoder reduces the rate to 10 kb/s.

2.4 By using both linear and logarithmic scales, plot the received power in watts and dBW of a line-of-sight link over distances from 100 m to 10 km under the assumptions of 2 GHz carrier frequency, 3 dBi transmit antenna, 13 dBi receive antenna, and 5 mW transmit power.

2.5 The coldest recorded temperature on Earth in a natural environment is about -90 °C. At what frequency is the standard $k_B T$ for thermal noise power spectral density approximation wrong by 10 percent?

2.6 By using the standard thermal noise power spectral density approximation, plot the noise power in units of dBm for a temperature of 20 °C for bandwidths from 1 kHz to 1 GHz.

2.7 For a system with 15 dBi antennas at both ends of the link that is operating at a carrier frequency of 28 GHz, a bandwidth of 100 MHz, a temperature of 300 K, and a transmit power of 1 mW, plot the SNR on a dB scale as a function of range for distances from 100 m to 100 km.

2.8 It is often useful to make quick estimates of receiver SNR. One should remember that $k_B T$ is about -204 dBW/Hz at 300 K. $(4\pi)^2 \approx 22$ dB. For a line-of-sight link,

SNR at the receiver is approximately

$$\text{SNR} = P_{tx} \frac{G_{tx} G_{rx}}{(4\pi)^2 (r_\lambda)^2 k_B T B}$$

$$\text{SNR (dB)} = P_{tx}(\text{dB}) + G_{tx} (\text{dB}) + G_{rx} (\text{dB})$$

$$- 22 - [-204 \text{ (dBW/Hz)} + B \text{ (dB)} + 2r_\lambda(\text{dB})]$$

$$= P_{tx}(\text{dB}) + G_{tx} (\text{dB}) + G_{rx} (\text{dB})$$

$$+ 182 \text{ (dBW/Hz)} - B \text{ (dB)} - 2r_\lambda(\text{dB}),$$

where P_{tx} is the transmit power; G_{tx} and G_{rx} are the transmit and receive gains, respectively; and r_λ is the link length in units of wavelengths. By using this form and assuming transmit and receive gains are 0 dB, quickly evaluate the received SNR for
(a) $P_{tx} = 1$ mW, $B = 10$ MHz, and $r_\lambda = 10,000$;
(b) $P_{tx} = 1$ W, $B = 1$ MHz, and $r_\lambda = 10^6$; and
(c) $P_{tx} = 1\,\mu$W, $B = 10$ MHz, and $r_\lambda = 10$.

2.9 By exploring external references, determine reasonable system parameters (power, gains, bandwidths, ranges) and resulting SNRs under a line-of-sight assumption for systems that operate as
(a) ground to geostationary satellite communications link;
(b) ground to low-Earth-orbit satellite communications link;
(c) local WiFi router; and
(d) body area network for smart watch to phone.

2.10 For a random variable that is drawn from a circularly symmetric complex Gaussian distribution of zero mean and unit variance, evaluate the probability that the magnitude of a particular draw from this distribution exceeds some value radius r.

3 Fundamental Limits on Communications

In this chapter, we discuss the fundamental limit – the Shannon limit – on data rate for a communications link. We motivate this limit by providing a sketch of the derivation. To construct this sketch, we discuss the idea of the ratio of hypersphere volumes and how the radius of Gaussian vectors converges to a known radius as the dimensionality goes to infinity. We provide a second approach to think about the Shannon limit, or equivalently channel capacity, of a data link by considering mutual information and entropy. By using a simple line-of-sight channel, we discuss the resulting link theoretical capacity and an approach to estimating the practical limit on data rate. Finally, we provide an overview of source coding approaches.

3.1 Information

Arguably, the most significant single contribution to communications theory within the last 100 years is the development of information theory, in particular Claude Shannon's efforts reported in the 1948 paper entitled "A Mathematical Theory of Communication" [7] and later in "Communication in the Presence of Noise" [8]. Information theory allows us to understand how much information can be moved through a channel. Actually, it could also be the single most important contribution to engineering in the last 100 years, but that is a crowded field.

Let us consider the simple additive Gaussian noise channel with signals at complex baseband for a communication system with one transmit antenna and one receive antenna. The mth sample in a sequence of received signal samples is indicated by z_m. The transmitted signal is indicated by s_m and noise is indicated by n_m. These signals are drawn from an independent and identically distributed (i.i.d.) zero-mean complex circularly symmetric Gaussian distribution. This distribution is natural for noise because of the central limit theorem. We typically use non-Gaussian distributions for our signaling, which we discuss in Chapter 6. However, this distribution maximizes the communications performance limit and is a reasonable approximation under many scenarios. Complex attenuation between the transmitter and receiver is indicated by a. Under this model, the transmit and receive signals are related by

$$z_m = a\, s_m + n_m. \tag{3.1}$$

In real life, channels are far more complicated; however, we can often relate those channels back to this one through some transformation or at least approximate the system performance with this model. The noise power is given by the variance of the noise. Often we assume units of power such that this variance is 1: $E\{\|n_m\|^2\} = 1$. In these units, we assume that power from the transmitter is given by $E\{\|s_m\|^2\} = P_s$. The resulting signal-to-noise ratio (SNR) is then given by

$$\text{SNR} = \frac{E\{\|z_m\|^2 - \|n_m\|^2\}}{E\{\|n_m\|^2\}}$$

$$= \frac{\|a\|^2 \ E\{\|s_m\|^2\}}{E\{\|n_m\|^2\}}$$

$$= \|a\|^2 P_s, \tag{3.2}$$

where we have assumed $E\{\|n_m\|^2\} = 1$.

With this model, we can say that the best-case c data transfer per complex channel (b/s/Hz) – or equivalently, (b/[s Hz]) – use is given by

$$c = \log_2(1 + \text{SNR}) = \log_2(1 + \|a\|^2 P_s). \tag{3.3}$$

This result, which we call the Shannon limit, can be interpreted to be the bound on the spectral efficiency. Spectral efficiency is the ratio of the data rate to the employed bandwidth. The units that we use for spectral efficiency are a little strange because (s Hz) is actually unitless. However, the form (b/s/Hz) suggests how to use spectral efficiency. The relationship defined in Equation (3.3) is depicted in Figure 3.1. At complex baseband, the maximum rate of independent samples would occur at B, which is the total bandwidth of the signal (including both positive and negative frequencies). Consequently, the upper bound on the achieved data rate C (b/s) is given by

$$C = Bc$$

$$= B \log_2(1 + \text{SNR}) = B \log_2(1 + \|a\|^2 P_s). \tag{3.4}$$

This result is not only an upper bound on performance but also theoretically achievable, although in practice we can only approach it. We often call this achievable limiting performance the *capacity* of the communications link or the channel. We also call this the Shannon bound, limit, or capacity. Because it is important, we have multiple names for it. The power of this result should not be underestimated. With a characterization of the power and propagation, one can determine the limiting data rate. From our perspective, this performance bound is remarkably useful as a system design tool.

As an example, we consider a system for which the received signal SNR is 10 dB for a signal that has a total bandwidth of 10 MHz. We assume that we have good filters, so noise outside our bandwidth of interest is not an issue. Within our band, we can have other problems. Our electronics can add additional noise. For example, let us assume that they double the noise power (that is, a noise figure of 3 dB, which is discussed in Section 4.3). The SNR at the receiver after the electronics is then approximately 10 dB − 3 dB = 7 dB. On a linear scale, the SNR is 7 dB ≈ 5. The corresponding maximum data rate is given by

$$C = 10 \cdot 10^6 \ \log_2(1 + 5) \approx 10 \cdot 10^6 \cdot 2.6 = 26 \,\text{Mb/s} \,. \tag{3.5}$$

In practice, because of the practical limits of coding and other issues, the actual data rate will be lower – maybe half of this rate, depending upon the error correcting code – but this does give an expectation of possible performance for a well-designed and -executed system.

When we operate in low-SNR environments, we can make a useful approximation of the capacity. The channel capacity can be approximated by

$$c = \log_2(1 + \text{SNR}) = \frac{\log_e(1 + \text{SNR})}{\log_e(2)}$$

$$\approx \frac{\text{SNR}}{\log_e(2)} \approx 1.44\,\text{SNR}, \tag{3.6}$$

for SNR $\ll 1$.

While it is typically useful for us to focus on complex channels (that are created by moving real passband signals to the complex baseband), on occasion we have problems defined by real channels. The SNR must be calculated for this real channel, and there is only one degree of freedom per use. Consequently, for a real channel, the capacity is given by

$$c = \frac{1}{2}\log_2(1 + \text{SNR}). \tag{3.7}$$

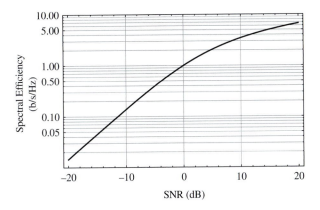

Figure 3.1 Shannon limit on channel capacity in terms of the spectral efficiency (b/s/Hz) as a function of SNR under the assumption of a complex channel.

3.2   Channel Capacity Motivation

While we will not develop a thorough formulation of the channel capacity because we are not careful in controlling how we approach infinity, we provide a motivational overview here. First, let us consider a set of sequences of channel uses. Each use of the channel is associated with a complex number, so the sequence of channel uses is given by a vector. At the receiver, each channel use is corrupted by noise. Consequently, the vector of the observed channel uses is given by a scaled version of the transmit vector plus a random vector of noise. One can think of this geometrically as there being a ball of noise around the scaled intended use vector. In some sense, the best we can do is to pack the intended channel uses so that the statistical balls of noise do not touch. We could then unambiguously determine which sequence was sent, as we see in Figure 3.2. The number of distinguishable sequences could be bounded by the ratio of the hypervolume of the observation hypersphere to the volume of the noise hypersphere. This heuristic motivation does not quite work because our balls of noise have unlimited extent, that is, the probability of noise exceeding some radius gets small quickly but never goes to zero. However, if we let the number of elements in the sequence go to infinity, then an interesting effect occurs, which is sometimes called *sphere hardening*. As the dimension of the vector space increases toward infinity, the fluctuations in the noise vector length decrease to the point where the noise vector approaches a constant-radius ball.

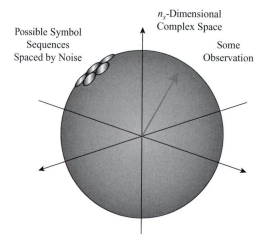

Figure 3.2 Notional depiction of some sequence of observations in an n_s-dimensional sample space separated by hyperspheres of noise.

To sketch the derivation of the capacity bound, let us first consider the volume of a hypersphere. For some radius r with m complex dimensions,[1] the hypervolume is given by

$$V(m, r) = \frac{\pi^m}{m!}(r^2)^m.$$ (3.8)

---

[1] This is as opposed to real dimensions, for which the hypervolume is given by

$$V_r = \frac{\pi^{n/2}}{\Gamma(n/2 + 1)} r^n.$$

For the ball of Gaussian noise, it would be tempting to replace r^2 with the variance of the noise in each observation σ_n^2. However, as we mentioned previously, the complex n_s-dimensional statistical ball of noise would fluctuate to a different radius if one were to draw a different sequence of observations given the same transmitted sequence. Here, we can invoke the concept of sphere hardening. In particular, as the dimensionality increases, the likelihood of a significant deviation from the surface of the hypersphere for the Gaussian ball decreases. If the vector of draw of the n-dimensional Gaussian vector is given by $\mathbf{g}$, then the radius squared of that draw is given by $x = r^2 = \|\mathbf{g}\|^2$. The distribution of the magnitude squared of n i.i.d. complex circularly symmetric Gaussian variables is given by the central complex χ^2 distribution that is discussed in Section 14.1.10. To keep the mean value of x the same, we scale the variance of each element in $\mathbf{g}$ to be $1/n$. The result of this discussion is that as n goes to infinity, the width of the distribution tends to zero, as is seen in Figure 3.3. Consequently, we can replace r^2 with σ^2 in this limit.

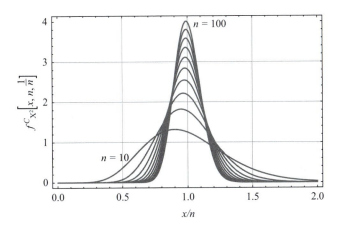

Figure 3.3 For an n-vector $\mathbf{g} \in \mathbb{C}^{n \times 1}$ with elements drawn independently from a circularly symmetric complex Gaussian distribution, the distribution of the random variable $x = \|\mathbf{g}\|^2$ is given by the complex χ^2 distribution $f_{\chi^2}^C[x, n, 1/n]$ with the variance of each element in $\mathbf{g}$ given by $1/n$. Probability density functions are displayed for dimensions given by $n = 10, 20, 40, \cdots, 100$.

We return to the discussion of the ratio of hypervolumes. Let us imagine that our transmitted sequence is also drawn from a Gaussian distribution, then, in the limit of a large number of channel uses (or symbols) n_s, we can directly determine the ratio. The hypervolume of the noise is given by

$$V_{n_s} = \frac{\pi^{n_s}}{n_s!}(\sigma_n^2 \, n_s)^{n_s}. \tag{3.9}$$

Because the variance of the observation is given by the sum of the variances of the received signal power P_r and the noise power σ_n^2, the hypervolume at the receiver is given by

$$V_r = \frac{\pi^{n_s}}{n_s!}([\sigma_n^2 + P_r]\, n_s)^{n_s}. \tag{3.10}$$

Thus, the bound on the number of different sequences n_{seq} (and note n_{seq} is typically not the same as n_s) that can be sent with n_s complex channel uses is given by the ratio of the volumes. The ratio is then given by

$$
\begin{aligned}
n_{\text{seq}} &= \frac{V_r}{V_n} \\
&= \frac{\frac{\pi^{n_s}}{n_s!}([\sigma_n^2 + P_r]\, n_s)^{n_s}}{\frac{\pi^{n_s}}{n_s!}(\sigma_n^2\, n_s)^{n_s}} \\
&= \left(\frac{\sigma_n^2 + P_r}{\sigma_n^2}\right)^{n_s} \\
&= \left(1 + \frac{P_r}{\sigma_n^2}\right)^{n_s}. \tag{3.11}
\end{aligned}
$$

The number of bits it takes to encode n_b is given by the base-2 logarithm of the number of different sequences n_{seq}, which is given by

$$
\begin{aligned}
n_b &= \log_2\left[\left(1 + \frac{P_r}{\sigma_n^2}\right)^{n_s}\right] \\
&= n_s \log_2\left(1 + \frac{P_r}{\sigma_n^2}\right). \tag{3.12}
\end{aligned}
$$

The bound on the bits per channel use c is given by

$$
\begin{aligned}
c &= \frac{n_b}{n_s} \\
&= \log_2\left(1 + \frac{P_r}{\sigma_n^2}\right) \\
&= \log_2(1 + \text{SNR}), \tag{3.13}
\end{aligned}
$$

where we observe that the definition of SNR is the ratio of received signal power to noise power. While our argument was not particularly precise, it turns out that this is the precise best-case bound in the limit of $n_s \to \infty$. Even more remarkable is that this bound is theoretically achievable, so this is the inner and outer bound of channel capacity.

3.2.1   Entropy Formulation of Channel Capacity

The capacity of a channel can also be described by the mutual information of the channel, which can be formulated in terms of entropy functions. The entropy of a random variable can be explained as the average number of bits required to encode its state. The entropy of a coin toss is 1 bit per toss. The entropy of a toss of a die is $\log_2(6) \approx 2.6$ bits per roll. It is slightly more complicated when the probabilities of various options are not equal. For a discrete random variable X with values $\{x_1, x_2, \cdots, x_M\}$, the entropy $H(X)$ is given by

$$H(X) = \sum_{m=1}^{M} p_{x_m} \log_2\left(\frac{1}{p_{x_m}}\right), \tag{3.14}$$

where p_{x_m} is the probability of the value x_m occurring. For two random variables X, with values $\{x_1, x_2, \cdots, x_M\}$, and Y, with values $\{y_1, y_2, \cdots, y_M\}$, the conditional entropy $H(Y|X)$ of random variable Y, given knowledge of the state of X, is given by

$$H(Y|X) = \sum_{m=1}^{M} \sum_{n=1}^{N} p_{x_m,y_n} \log_2\left(\frac{1}{p_{y_n|x_m}}\right)$$

$$= \sum_{m=1}^{M} \sum_{n=1}^{N} p_{x_m,y_n} \log_2\left(\frac{p_{x_m}}{p_{x_m,y_n}}\right). \tag{3.15}$$

Here, p_{x_m,y_n} indicates the joint probability and $p_{y_n|x_m}$ indicates the conditional probability of y_n given x_m. We employed the relationship

$$p_{x_m,y_n} = p_{x_m} \, p_{y_n|x_m}. \tag{3.16}$$

It is common in communications to consider the potential of continuous random variables. In this scenario, which implies continuous random variable X, we use differential entropy $h(X)$ given by

$$h(X) = \int dx \, p(x) \log_2\left(\frac{1}{p(x)}\right), \tag{3.17}$$

where $p(x)$ is the probability density function for the random variable X. Similar to the discrete case, the conditional differential entropy is given by

$$h(Y|X) = \int dx \, dy \, p(x, y) \log_2\left(\frac{p(x)}{p(x, y)}\right). \tag{3.18}$$

Let's consider a very simple channel of real zero-mean continuous random Gaussian signal X being received in the presence of additive real zero-mean random Gaussian noise W. The random variable associated with the received signal Y is then given by

$$Y = X + W. \tag{3.19}$$

We are interested in the mutual information (which is the capacity when the mutual information is maximized) between X and Y in the presence of this distortion. The mutual information is given by

$$c = \max_{p(X)} I(X; Y)$$

$$= \max_{p(X)} \int d^n y \int d^m x \, p(x, y) \log_2\left(\frac{p(x, y)}{p(x) \, p(y)}\right) \tag{3.20}$$

$$= \int d^n y \int d^m x \, p(x, y) \left[\log_2\left(\frac{1}{p(y)}\right) + \log_2\left(\frac{p(x, y)}{p(x)}\right)\right]$$

$$= \int d^n y \, p(y) \, \log_2 \left(\frac{1}{p(y)} \right) - \int d^n y \int d^m x \, p(x, y) \, \log_2 \left(\frac{p(x)}{p(x, y)} \right)$$

$$= \max_{p(X)} \, [h(Y) - h(Y|X)], \tag{3.21}$$

where we use $d^n y$ and $d^m x$ when the random variable is a higher-dimensional object. For example, if X and Y are complex variables, $d^2 x = d\Re x \, d\Im x$ and $d^2 y = d\Re y \, d\Im y$. We have used the observation that

$$\int dx \, dy \, p(x, y) f(y) = \int dy \, p(y) f(y). \tag{3.22}$$

The observed signal Y is the sum of two zero-mean uncorrelated Gaussians, so Y is zero mean and has the variance of the sum of the variances of X and W so that

$$\sigma_Y^2 = \sigma_X^2 + \sigma_W^2. \tag{3.23}$$

The entropy of a real Gaussian random variable Y with variance σ_Y^2 is given by

$$h(Y) = \log_2 \left(\sqrt{2\pi \, \sigma_Y^2 \, e} \right). \tag{3.24}$$

The conditional entropy first appears more complicated until one realizes that the entropy of Y given X is just the entropy of the noise W, so

$$h(Y|X) = h(W) = \log_2 \left(\sqrt{2\pi \, \sigma_W^2 \, e} \right). \tag{3.25}$$

Thus, the mutual information or channel capacity is given by

$$I(X; Y) = h(Y) - h(W)$$

$$= \log_2 \left(\sqrt{2\pi \, \sigma_Y^2 \, e} \right) - \log_2 \left(\sqrt{2\pi \, \sigma_W^2 \, e} \right)$$

$$= \log_2 \left(\sqrt{2\pi \, [\sigma_X^2 + \sigma_W^2] \, e} \right) - \log_2 \left(\sqrt{2\pi \, \sigma_W^2 \, e} \right)$$

$$= \log_2 \left(\frac{\sqrt{2\pi \, [\sigma_X^2 + \sigma_W^2] \, e}}{\sqrt{2\pi \, \sigma_W^2 \, e}} \right)$$

$$= \log_2 \left(\sqrt{\frac{2\pi \, [\sigma_X^2 + \sigma_W^2] \, e}{2\pi \, \sigma_W^2 \, e}} \right)$$

$$= \log_2 \left(\sqrt{\frac{\sigma_X^2 + \sigma_W^2}{\sigma_W^2}} \right)$$

$$= \log_2 \left(\sqrt{1 + \frac{\sigma_X^2}{\sigma_W^2}} \right)$$

$$= \frac{1}{2} \log_2 \left(1 + \frac{\sigma_X^2}{\sigma_W^2} \right). \tag{3.26}$$

The term σ_X^2/σ_W^2 is simply the SNR of the received signal.

For modern communications, it is often useful to consider complex baseband formulations, so we re-evaluate the above calculation given a complex Gaussian. The entropy of a circularly symmetric complex Gaussian random variable Y with variance σ_Y^2 is given by

$$h(Y) = \log_2\left(\pi \, \sigma_Y^2 \, e\right). \tag{3.27}$$

The resulting mutual information is then given by

$$c = I(X; Y) = h(Y) - h(W)$$
$$= \log_2(2\pi \, \sigma_Y^2 \, e) - \log_2(\pi \, \sigma_W^2 \, e)$$
$$= \log_2\left(1 + \frac{\sigma_X^2}{\sigma_W^2}\right). \tag{3.28}$$

Once again, the term σ_X^2/σ_W^2 is the SNR of the received signal. This result corresponds to the channel capacity determined in Equation (3.3).

3.2.2   Energy per Bit

It is sometimes useful to think about the energy per information bit normalized by the noise spectral density E_b/N_0. This ratio is sometimes informally called "eb-no." The energy per information bit E_b (which is often not the same as symbol bit because of error correcting codes or spreading) is given by

$$E_b = \frac{P_r}{R}, \tag{3.29}$$

where R is the data rate in units of bits per second. The noise spectral density is indicated by N_0, so the total noise power is given by $\sigma_W^2 = N_0\,B$, where B is the total bandwidth.

We can express the Shannon limit in terms of E_b/N_0, which leads to an interesting bound in the limit of low spectral efficiency. If we assume that the communications link is operating at channel capacity, then $R/B = c$. By using this information, we express the channel capacity for spectral efficiency in terms of

$$c = \log_2(1 + \text{SNR}) = \log_2\left(1 + \frac{P_r}{\sigma_W^2}\right)$$
$$= \log_2\left(1 + \frac{E_b\,R}{N_0\,B}\right)$$
$$= \log_2\left(1 + \frac{E_b}{N_0}\frac{R}{B}\right)$$
$$= \log_2\left(1 + \frac{E_b}{N_0}c\right). \tag{3.30}$$

The relationship between capacity and E_b/N_0 is displayed in Figure 3.4. In this form the capacity is expressed implicitly as a transcendental relationship. However, if we assume that the spectral efficiency is small, then we can expand the logarithm by using

$$c = \log_2 \left(1 + \frac{E_b}{N_0} c \right)$$

$$= \log_2(e) \log \left(1 + \frac{E_b}{N_0} c \right)$$

$$\approx \log_2(e) \frac{E_b}{N_0} c \quad ; \quad \text{for small } \frac{E_b}{N_0} c$$

$$1 \approx \log_2(e) \frac{E_b}{N_0}$$

$$\frac{E_b}{N_0} \approx \frac{1}{\log_2(e)} \approx -1.6 \text{ dB} . \tag{3.31}$$

The implication of this result is that -1.6 dB is the minimum ratio of energy per information bit to noise spectral density E_b/N_0 required to produce a theoretically viable communications link. This bound on E_b/N_0 is only valid at very low spectral efficiency. Larger values of E_b/N_0 are required at higher spectral efficiencies.

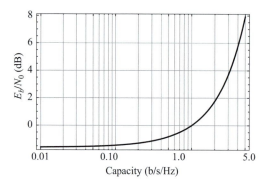

Figure 3.4 The best-case required E_b/N_0 to achieve a given communications spectral efficiency.

3.3   Estimating Simple System Performance

By combining concepts from Sections 3.1 and 2.2.2 on the information theoretic limits and propagation attenuation, respectively, we can provide some simple results on expected communications rates. From the Friis equation (Equation 2.5), we know that line-of-sight received power is given by

$$P_r = \frac{G_t G_r}{(4\pi r/\lambda)^2} P_t, \tag{3.32}$$

where P_t is the power of the transmitter; G_t and G_r are the transmit and receive antenna gains, respectively; r is the distance between the transmit and receive antennas; and λ is the carrier wavelength. From Equation (2.12), we also know that the noise power is given by

$$P_{\text{noise}} = k_B T B, \tag{3.33}$$

where k_B is the Boltzmann constant, T is the absolute temperature, and B is the full bandwidth, so that the SNR is given by

$$\text{SNR} = \frac{P_r}{P_{\text{noise}}}$$

$$= \frac{G_t G_r}{(4\pi \, r/\lambda)^2 \, k_B \, T \, B} P_t. \tag{3.34}$$

The capacity in terms of data rate (b/s) under the assumption of complex baseband signal is given by

$$C = B \log_2(1 + \text{SNR})$$

$$= B \log_2 \left(1 + \frac{G_t G_r}{(4\pi \, r/\lambda)^2 \, k_B \, T \, B} P_t \right). \tag{3.35}$$

3.3.1   Degraded Performance

For a variety of reasons, the performance indicated by Equation (3.35) is not achievable. Real electronics increase the noise observed in the receiver. This is typically parameterized by the noise figure. Real error correcting codes cannot quite achieve the theoretic capacity. Finally, the full bandwidth allocation cannot be completely used because the spectral shape cannot be perfectly confined, so spectral guard bands are typically employed. We can characterize a modified capacity result that approximates a more realistic limit on the achievable data rate with the form

$$C_{\text{eff}} = \alpha \, B \, c_{\text{eff}}$$

$$c_{\text{eff}} = \log_2(1 + \text{SNR} \, \beta)$$

$$= \alpha B \log_2 \left(1 + \beta \frac{G_t G_r}{(4\pi \, r/\lambda)^2 \, k_B \, T \, B} P_t \right), \tag{3.36}$$

where $0 < \alpha \leq 1$ introduces that effect of imperfect use of a spectral allocation and $0 < \beta \leq 1$ incorporates the effects of imperfect error correction codes of the noise figure of the electronics (as discussed in Section 4.3). While every system is different, it is not uncommon for the spectral degradation parameter α to have a value of 0.8 to 0.9 and for the combined noise figure and error correction degradation parameter β to have values between $1/20$ (or -13 dB) and $1/4$ (or -6 dB). Higher-performance systems have values closer to 1 (or 0 dB).

There is also another limiting effect: at very high SNR, the theoretical effective spectral efficiency c_{eff} can be quite high, even after degrading the performance. Because of other system effects, such as local oscillator phase noise and even transmitter noise, very high spectral efficiency modulations are not achievable in practice. As will be discussed in Chapter 6, high-order modulations (or equivalently large constellations) are required to get to high spectral efficiencies. It is more difficult to work with these high-order modulations. Although there are extremely high-performance systems that can achieve spectral efficiencies of up to 9 b/s/Hz, most systems are limited to spectral efficiencies of less than 5 b/s/Hz, and many cannot achieve more than 3 b/s/Hz. Consequently, if the effective bound on spectral efficiency exceeds this number, then we need to supplant the spectral efficiency with a constrained value.

3.3.2   Across-Room Performance

For an example, let us consider a common, modern, in-room WiFi scenario. We assume that it is line-of-sight. Let us also assume that the antennas are isotropic with 0 dBi gain, the carrier frequency is 5.8 GHz, the bandwidth is 20 MHz, the range is 4 m, and the transmit power is 10 mW. The first step is to calculate the wavelength, which is given by

$$\lambda = \frac{c}{f_c}$$

$$\approx \frac{3 \cdot 10^8 \text{ m/s}}{5.8 \cdot 10^9 \text{ Hz}} = 0.052 \text{ m}. \tag{3.37}$$

Let us assume, because this is a relatively inexpensive system, degradation parameters of $\alpha = 0.8$ and $\beta = 1/10$. The approximate limit on expected performance is then given by

$$c_{\text{eff}} = \log_2\left(1 + \beta \, \frac{G_t \, G_r}{(4\pi \, r/\lambda)^2 \, k_B \, T \, B} \, P_t\right)$$

$$= \log_2\left(1 + \frac{(1/10)\,(1)\,(1)\,(10 \cdot 10^{-3} \text{ W})}{(4\pi \, 4/0.052)^2 \, (1.38 \cdot 10^{-23} \text{ J/K})\,(300 \text{ K})\,(20 \cdot 10^6 \text{ Hz})}\right)$$

$$\approx 14.0 \text{ b/s/Hz}. \tag{3.38}$$

As was discussed in Section 3.3.1, this spectral efficiency is not practically achievable. Consequently, we will limit it to $\bar{c}_{\text{eff}} = 5$ b/s/Hz. The resulting data rate is given by

$$C_{\text{eff}} = \alpha \, B \, \bar{c}_{\text{eff}}$$

$$= (0.8)\,(20 \cdot 10^6 \text{ Hz})\, 5 \text{ b/s/Hz}$$

$$\approx 80 \text{ Mb/s}. \tag{3.39}$$

3.3.3   Through-the-Wall Performance

Let us reconsider the previous problem but with the addition of wall loss. Imagine that the signal has both line-of-sight propagation and two walls through which it must pass. Attenuation through walls is a strong function of carrier frequency and wall material, so there is no one loss number. However, at 5.8 GHz and normal interior wall construction, a loss of 20 dB per wall is not unreasonable. For two successive walls, the effect multiplies (or adds on a logarithmic scale), so the additional effect is 40 dB more loss (or 10^4 on a linear scale). Consequently, we multiply our original SNR approximation by $l^2_{\text{wall}} = 10^{-4} = 2 \cdot 20 \text{ dB} = 40 \text{ dB}$ for two walls. The effective limit on spectral efficiency becomes

$$c_{\text{eff}} = \log_2\left(1 + \beta \, \frac{G_t \, G_r}{(4\pi \, r/\lambda)^2 \, k_B \, T \, B} \, P_t \, l^2_{\text{wall}}\right)$$

$$= \log_2\left(1 + \frac{(1/10)\,(1)\,(1)\,(10 \cdot 10^{-3} \text{ W})\,(10^{-4})}{(4\pi \, 4/0.052)^2 \,(1.38 \cdot 10^{-23} \text{ J/K})\,(300 \text{ K})\,(20 \cdot 10^6 \text{ Hz})}\right)$$

$$\approx 1.2 \text{ b/s/Hz}. \tag{3.40}$$

The corresponding limit on data rate is then given by

$$C_{\text{eff}} = \alpha \, B \, c_{\text{eff}}$$

$$= (0.8)\,(20 \cdot 10^6 \text{ Hz})\,1.2 \text{ b/s/Hz}$$

$$\approx 19 \text{ Mb/s}. \tag{3.41}$$

3.3.4 Satellite Performance

We depict a notional geostationary satellite-to-ground communications link in Figure 3.5. Because of the cost of these systems, they are designed carefully. Here, let us consider the data rate from a notional design. Geostationary satellites are used because they are approximately stationary as viewed from the rotating Earth. As Earth rotates, the satellite's orbit follows along. Consequently, it is much easier to use antennas with higher gain. Once you have the antenna pointed correctly, the direction should be constant.

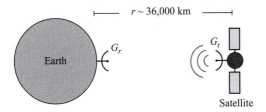

Figure 3.5 Notional satellite-to-ground communications link.

For this example, let us assume that the ground antenna has a gain of 23 dBi (or 200 on a linear scale), the satellite antenna has a gain of 30 dBi (or 1000 on a linear scale), the carrier frequency is 10 GHz, the bandwidth is 1 MHz, the range is 36,000 km, and the transmit power is 10 W. The wavelength is approximately $\lambda \approx 0.03$ m. For this example we assume degradation parameters of $\alpha = 0.9$ and $\beta = 1/4$ because the system is well designed. However, there is the prospect of additional attenuation loss because of the effects of the atmosphere and, in particular, rain. At higher frequencies, these effects can be quite significant, but we can assume that the atmospheric loss is given by $l_{\text{atmos}} = -3$ dB (or $1/2$ on a linear scale) for our example:

$$
\begin{aligned}
c_{\text{eff}} &= \log_2\left(1 + \beta \, \frac{G_t \, G_r}{(4\pi \, r/\lambda)^2 \, k_B \, T \, B} \, P_t \, l_{\text{atmos}}\right) \\
&= \log_2\left(1 + \frac{(1/4)\,(200)\,(1000)\,(10 \text{ W})\,(1/2)}{(4\pi \, 3.6 \cdot 10^7 \text{m}/0.03\text{m})^2 \,(1.38 \cdot 10^{-23} \text{ J/K})\,(300 \text{ K})\,(10^6 \text{ Hz})}\right) \\
&\approx 0.34 \, \text{b/s/Hz}.
\end{aligned}
\tag{3.42}
$$

The corresponding limit on data rate is then given by

$$
\begin{aligned}
C_{\text{eff}} &= \alpha \, B \, c_{\text{eff}} \\
&= (0.9)\,(1 \cdot 10^6 \text{ Hz})\,0.34 \, \text{b/s/Hz} \\
&\approx 0.3 \, \text{Mb/s}.
\end{aligned}
\tag{3.43}
$$

3.4   Sources

Digital communications systems can send any digital representation of information, as indicated in Figure 3.6. In many cases, the sources of information are originally analog signals, such as voice or a visual image. These signals are then converted to digital signals via an analog-to-digital conversion process. Once we have a sequence of bits, our digital communications system can move that information to the receiver. However, we have skipped an important step in this discussion. Typically, we can actually reduce the number of bits required to represent this information. This operation is called *source compression*. It is generally undesirable to send uncompressed data because extra bits are required to send the same useful information. In addition, good source coding or compression ensures that the ones and zeros of the signal are equally likely and removes correlations from the sequence of bits.

3.4.1   Source Compression and Coding

How many bits do you need to represent information without any loss? Imagine there were eight options that could occur with equal probability. It would then require

$$
\log_2(8) = 3 \, \text{bits}
\tag{3.44}
$$

to represent which of the eight options was the correct one. The bit sequence required to specify each choice must be different; otherwise, it just does not work.

The solution space gets stranger if the probability of each option is not the same. It is probably not surprising to suggest that if an outcome is unexpected, then this outcome is surprising. To save bits, you might determine how to reduce the number of bits required to identify an outcome that is likely and allow for longer names (more bits) for relatively rare occurrences. Imagine we consider a scenario with four options, $\log_2(4) = 2$ bits. Let us assume that the probability distribution for these

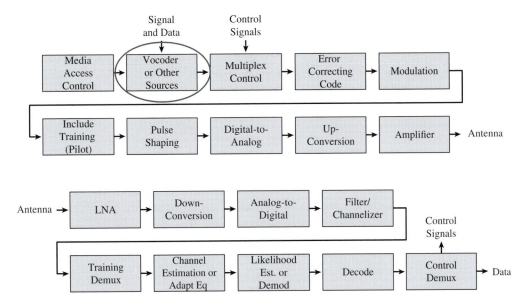

Figure 3.6 Source of information for communications systems.

options is given by $p_1 = 1/2, p_2 = 1/4, p_3 = p_5 = 1/8$. The naive approach requires 2 bits per choice (Table 3.1).

Option	Probability	Bit label
1	1/2	00
2	1/4	01
3	1/8	10
4	1/8	11

Let us build some potential labels that require different numbers of bits. If we require the length of the sequence to be expressed in the coding, then the labeling shown in Table 3.2 would work.

Option	Probability	Bit label
1	1/2	0
2	1/4	10
3	1/8	110
4	1/8	111

The average number of bits required is $7/4 < 2$ bits, so we have reduced the number of bits required to specify which option is selected. Thus, we have lossless source coding with a reduced number of bits. If a sequence of sections is being transmitted, then this particular encoding has a secondary advantage. The choice can be decoded immediately. If the code for option 1 is sent (bit 0), then another bit 0 indicates another option 1. If the next bit is a 1, then the decoder needs to look at more bits. In each case, it is clear when a new code word has started. This code-word-by-code-word decodability may not be a requirement because it is equally valid to only be able to decode at the end of a long sequence of code words, but there certainly is practical benefit to being able to decode immediately.

3.4.2 Lossless Source Compression and Coding

Given the example in Section 3.4.1, it is reasonable to ask how well one can do on average. The answer is given by the entropy associated with the probability distribution of the underlying signal being compressed. The entropy for a random variable X with values $\{x_1, x_2, \cdots, x_N\}$ associated with probabilities p_{x_m} is given by

$$H(X) = \sum_{m=1}^{N} p_{x_m} \log_2\left(\frac{1}{p_{x_m}}\right). \tag{3.45}$$

For the example of eight equally weighted options,

$$H(X) = \sum_{m=1}^{8} p_{x_m} \log_2\left(\frac{1}{p_{x_m}}\right)$$

$$= \sum_{m=1}^{8} \frac{1}{8} \log_2(8) = 3 \text{ bits}, \tag{3.46}$$

which we already knew, but it is nice that the entropy evaluation provides the same answer. For the unequal probability example in Section 3.4.1,

$$H(X) = \sum_{m=1}^{4} p_{x_m} \log_2\left(\frac{1}{p_{x_m}}\right)$$

$$= \frac{1}{2} \log_2(2) + \frac{1}{4} \log_2(4) + 2\frac{1}{8} \log_2(8)$$

$$= \frac{1}{2} + \frac{2}{4} + 2\frac{3}{8} = \frac{7}{4} \text{ bits}, \tag{3.47}$$

which was achieved by the source coding represented in the example.

3.4.3 Lossy Source Compression

Often we do not need to maintain an exact representation of the information that we wish to communicate. For instance, we regularly compress audio, images, and videos so that they cannot be reconstructed exactly, but are good enough that we do not notice any errors caused by the compression. In cellular phones, the signal processing to this lossy compression for converting voice to a digital signal is called a vocoder. Lossy compression is so common in images (such as JPEGs) and videos (such as MPEGs) that we sometimes forget that it came from a much larger signal source.

The theoretical limits for performance are addressed by rate-distortion theory (see Reference [9] for a more thorough discussion). The basic formulation of mutual information between random variables X and Y, which we discuss in Section 3.2.1, is given by

$$I(Y;X) = h(Y) - h(Y|X), \tag{3.48}$$

where $h(Y)$ is the entropy of Y and $h(Y|X)$ is the conditional entropy of Y given that X is known, which reduces to the entropy of the noise for many traditional problems.

For rate distortion, the problem is more complicated because the lossy compression distortion increases the effective (if not real) noise. Rate-distortion theory formalizes the question: *What is the smallest number of bits that I have to send given that the expected distortion is no worse than some value?* Mathematically, for some worst-case expected distortion D, the smallest rate R is given by

$$R(D) \geq \min_{p(y|x) \,:\, D \geq \int dx\, dy\, p(y|x)\, d(x,y)} I(Y;X). \tag{3.49}$$

First, rate-distortion theory does not specify how to quantify this distortion; however, it is often specified as the mean squared error for problems that can be solved analytically. The rate-distortion optimization is then given by

$$R(D) \geq \min_{p(y|x) \,:\, D \geq E\{(x-y)^2\}} I(Y;X). \tag{3.50}$$

As one of the few analytically solvable problems, let us consider the real Gaussian source problem with a mean squared error distortion metric. The mutual information is given by

$$I(X;Y) = h(Y) - h(Y|X) = h(X) - h(X|Y)$$
$$= \frac{1}{2}\log_2(2\pi\,\sigma^2\,e) - h(X|Y), \tag{3.51}$$

where we have used not only the relationship in Equation (3.24) but also the observation that $h(X|Y) = h(X - Y|Y)$ because shifting the random variable by Y when Y is known does not change the entropy. Next, we observe that Gaussian distributions maximize entropy for a given variance, so the conditional entropy is given by

$$h(X|Y) = h(X - Y|Y) = \frac{1}{2}\log_2(2\pi\, E[(X - Y)^2]\, e). \tag{3.52}$$

To maximize this conditional entropy, we set $E[(X - Y)^2] = D$. The resulting mutual information $I(X;Y)$ is given by

$$I(X;Y) = h(Y) - h(Y|X) = h(X) - h(X|Y)$$
$$= \frac{1}{2}\log_2(2\pi\,\sigma^2\,e) - \frac{1}{2}\log_2(2\pi\,D\,e)$$
$$= \frac{1}{2}\log_2\left(\frac{\sigma^2}{D}\right)$$
$$R(D) \geq \frac{1}{2}\log_2\left(\frac{\sigma^2}{D}\right) \quad \forall\, \sigma^2 > D. \tag{3.53}$$

Consequently, the data rate $R(D)$ is given in terms of bits per sample as a function of distortion variance. We limit distortion to be less than the variance of the Gaussian signal; otherwise, rates become negative because distortion overwhelms the signal to the point that you are better off sending nothing. The achievable rates are above the line in Figure 3.7.

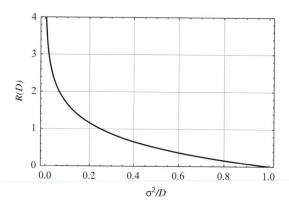

Figure 3.7 Rate-distortion limit for real Gaussian signal under a mean squared error distortion metric.

Problems

3.1 For complex symbols, evaluate numerically the maximum number of bits per symbol possible in the following cases:
(a) SNR = 0 dB
(b) SNR = 4 dB
(c) SNR = −10 dB
(d) SNR = −20 dB

3.2 For a system with an SNR of 6 dB and bandwidth of 1 MHz:
(a) Evaluate the bounding spectral efficiency in b/symbol for real symbols.
(b) Evaluate the bounding spectral efficiency in b/symbol for complex symbols.
(c) Evaluate the bounding data rate assuming complex symbols.

3.3 Consider the channel capacity for a wireless system with a signal received with a total bandwidth of 4 MHz and an SNR of −17 dB.
(a) Explain the meaning of the Shannon limit.
(b) Provide a numerical approximation and evaluate the best theoretical spectral efficiency.
(c) Evaluate the best data rate.
Remember to include units.

3.4 By assuming that a communications link is operating in the low spectral efficiency limit with complex symbols, evaluate the required received SNR for rates and bandwidths:

(a) $R = 1$ Mb/s, $B = 100$ MHz
(b) $R = 10$ b/s, $B = 10$ kHz
(c) $R = 1$ kb/s, $B = 50$ kHz

3.5 Consider a line-of-sight link. For a system with 15 dBi antennas at both ends of the link, and that is operating at a carrier frequency of 28 GHz, a bandwidth of 100 MHz, a temperature of 300 K, and a transmit power of 1 mW, plot the capacity as a function of range for distances from 100 m to 100 km.

3.6 For a communications system operating at small SNR, what is the approximate achievable rate as a function of bandwidth B and SNR?

3.7 Consider a received mth sample z_m of a complex baseband signal in the presence of additive complex circularly symmetric Gaussian noise n_m of the form

$$z_m = a\, s_m + n_m,$$

where s_m is the transmitted signal. We assume the effects of amplification are absorbed by the coefficient a, and that s_m has unit variance. Assume that the bandwidth of the signal is 1 MHz, the temperature is 300 K, and the SNR is 6 dB.
(a) Evaluate the value of $\|a\|^2$.
(b) By assuming that the phase of a is zero, evaluate the value of a.
(c) By assuming that even entries of s_m have a value of $+1$ and that odd entries have a value of -1, generate a scatter plot where each point corresponds to a draw of z_m in the complex domain with axis of $\Re\{z_m\}$ and $\Im\{z_m\}$ for 200 random draws of z_m.

3.8 For complex signals, evaluate the required SNR as a function of desired spectral efficiency, and plot the SNR on a linear scale for spectral efficiencies from 0.1 to 10.

3.9 Evaluate the entropy of a system that has four states (A, B, C, D) with the following sets of probabilities:
(a) 1/4, 1/4, 1/4, 1/4
(b) 1/2, 1/6, 1/6, 1/6
(c) 1/2, 1/4, 1/8, 1/8
(d) 3/4, 1/8, 1/16, 1/16

3.10 Evaluate the average required bits per sample required for lossless source coding for the following:
(a) 100 possible values that are equally likely;
(b) five values that have likelihoods fall symmetrically about the center state as a triangle with relative values of 1, 2, 3, 2, 1; and
(c) a set of eight possible values for which each successive value is half as likely as the previous one.

3.11 Consider the hypervolumes of a real hypercube of dimension n and length $2r$ on a side and a real hypersphere of the same dimension contained within this hypersphere of radius r.

(a) Evaluate the ratio of hypersphere to hypercube hypervolumes.

(b) Plot the ratio of the hypervolumes for dimensions 2 to 10.

3.12 For a random signal with each sample drawn independently from a real Gaussian distribution of variance 10, find the theoretically required number of bits per sample to achieve a lossy distortion variance of D given by

(a) 1

(b) 0.1

(c) 10^{-6}

(d) 10, explain

3.13 Plot and mathematically describe the ratio of the standard deviation to the mean of a complex central χ^2 distribution as the number of contributing complex Gaussian random variables increases.

3.14 Design and specify the requirements for a communications system at a center frequency of 10 GHz that sends a 1 b/s communications link to a planet orbiting the star nearest to our solar system, Alpha Centauri (4.37 ly). Specify antenna gain, approximate antenna size, power, and bandwidth.

4 Amplifiers and Noise

In this chapter, we discuss the concepts of amplification that we use to overcome noise. We review the idea of power amplifiers that are used by the transmitter and introduce metrics for nonlinear contributions in signal amplification. We discuss the concept of the low-noise amplifier (LNA) that is typically the first amplifier in the receiver chain. We motivate this type of amplifier by introducing the idea of the noise figure. Finally, we review the idea of automatic gain control.

4.1 Amplifiers

As we mentioned in Section 3.1, the fundamental limit in communications performance is determined by the signal-to-noise ratio (SNR). To increase this ratio, we can increase the signal power and decrease the noise power. In both cases, amplifiers can help. We indicate the location of these amplifiers in the transmitter and receiver processing chains in Figure 4.1.

For a wireless communications link, because the transmitted signal is typically attenuated significantly as it propagates through a channel, it is useful to employ an amplifier to increase the power at the transmit antenna. Consequently, the last major component in the transmit chain before the antenna is commonly a power amplifier.

Similarly, the first major component in the receive chain after the antenna is typically an LNA, although we often place a band-pass (pre-selection) filter before the amplifier and the antenna. We show in Section 4.3 that the performance of a receiver can be affected dramatically by the performance of the first amplifier in a chain of active components.

4.2 Power Amps

As you might expect from the name, the primary role of a power amplifier is to increase the output power at the transmitter. The signal at the input of the amplifier is multiplied by some factor, so for some input signal $x(t)$, the ideal output signal is given by $y(t)$ as a function of time t:

$$y(t) = \sqrt{g}\, x(t), \tag{4.1}$$

where g is the gain of the amplifier in terms of power, which is often expressed on a decibel scale.

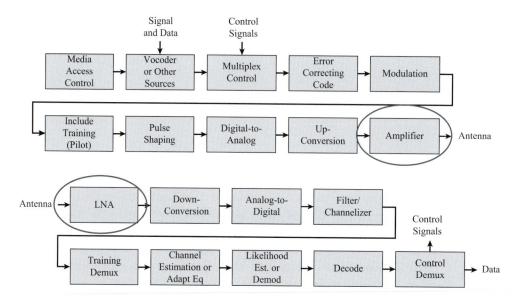

Figure 4.1 Connected to the radio's antennas, amplifiers are the last and first significant blocks in a radio's transmit and receive processing chains.

It is worth noting that active circuits add noise in addition to the thermal noise. This is a critical issue not only in the receiver, as we discuss in Section 4.3, but also in the transmitter. Because the intended signal is typically quite large compared to the noise added by the transmit amplifier, the transmitter noise effect is often ignored. However, this noise can be important for some systems. As an example, full-duplex systems transmit and receive simultaneously; thus, the transmitter noise can affect the receiver noise.

The range of transmitter output powers, associated with the *power amplifier*, is remarkable. A body area network radio may be able to close its link with a few hundred microwatts ($\sim 100\,\mu W$), while some long-range analog AM radio stations use megawatt transmitters ($\sim 1\,MW$), differing in power by 10 orders of magnitude. Clearly, the technologies of these amplifiers have little to do with each other, but they do serve similar functions. Because the range of power in communications systems varies widely, it is often convenient to express the power on a decibel scale. Two common units are dBW and dBm. For these two scales, the power is referenced to watts or milliwatts, respectively. For example, 100 W of transmitted power corresponds to 20 dBW = 50 dBm. As a common example, WiFi systems are typically limited to 100 mW = 20 dBm = -10 dBW.

While the largest power produced by solid state amplifiers keeps increasing, the really big amplifiers use some combinations of electron beams and resonant cavities in vacuum tubes or chambers. Two examples of this type of technology are the klystron and magnetron. Magnetrons are used to produce kilowatt power in microwave ovens.

In practice, there is often a trade-off in terms of the accuracy and the efficiency of the amplifier. While an accurate model of amplifier characteristics can be complicated (and is often described by employing Volterra series expansions), if we ignore any delay-dependent frequency selectivity, we can approximate the amplitude functionality with a Taylor series expansion, which is given by

$$y(t) = \sqrt{g} f(x(t))$$

$$f(x(t)) = \sum_{m=1}^{\infty} a_m (x(t))^m \tag{4.2}$$

$$a_1 = 1.$$

Because many amplifiers express some symmetry about the mean value of $x(t)$, often the odd values of a_m dominate the even values. Furthermore, the magnitude of the values of a_m fall quickly for a well-designed amplifier. We display examples of amplifier transfer functions in Figure 4.2. In the figure, we show both a linear amplifier with a power gain of $g = 20$ dB (or amplitude gain of 10) and a nonlinear amplifier (with $a_1 = 1$, $a_3 = 0.01$, and all other coefficients $= 0$ in Equation (4.2)) that deviates from the ideal amplifier at higher input voltages.

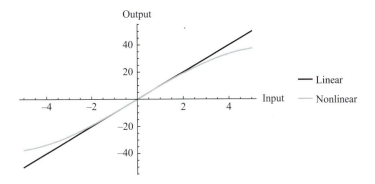

Figure 4.2 Relationship between input and output voltage for an idealized linear and nonlinear amplifier amplitude gain. Power gain of 20 dB.

There are multiple ways in which these amplifier nonlinearities are parameterized on data sheets. One common approach is to define the 1 dB compression point, which is the input or output power at which the output power is 1 dB lower than an ideal amplifier of the specified gain.

Another common parameterization approach is the intercept point (IP) or, more specifically, the input intercept point (IIP). This parameterization seems a little strange at first, so we will work through an explanation. Let us assume that the dominant nonlinearity is the third-order harmonic, which is commonly true. Our nonlinear equation relating the real input amplitude (not power) to the real output amplitude is given by

$$y(t) = a_1 x(t) - a_3 x^3(t),\tag{4.3}$$

where both a_1 and a_3 are real and positive in our example. Let us assume that $x(t)$ is constructed from a real tone:

$$x(t) = a_0 \cos(2\pi f).\tag{4.4}$$

At the output of the linear component of the amplifier, the tone is present at the correct frequency and its output is increased in amplitude by a_1, so the amplitude of the output tone is given by $a_0 a_1$. The nonlinear contribution has a different effect. We expand the cube of the tones, which is given by

$$a_3 [x(t)]^3 = a_3 [a_0 \cos(2\pi f)]^3$$

$$= \frac{a_0^3 a_3}{4}[3 \cos(2\pi ft) + \cos(2\pi [3f]t)].\tag{4.5}$$

A new tone is constructed at $3f$. The sum of the two components is given by

$$y(t) = a_1 a_0 \cos(2\pi f) - \frac{a_0^3 a_3}{4}[3 \cos(2\pi ft) + \cos(2\pi [3f]t)]$$

$$= \left[a_1 a_0 - \frac{3a_0^3 a_3}{4}\right]\cos(2\pi f) - \frac{a_0^3 a_3}{4}\cos(2\pi [3f]t).\tag{4.6}$$

If the input amplitude a_0 and the nonlinearity a_3 are small, then the linear effect is negligible. As we increase the input amplitude a_0, the nonlinear contributions increase quickly, as the third power of the input amplitude. While one would never operate in this regime, one can imagine increasing the input amplitude to the point that the linear contributions and the nonlinear contributions are equal. To define the intercept point, we ignore the new frequency and ask when the contributions are equal for the coefficient of $\cos(2\pi ft)$. The input amplitude at which this occurs defines the intercept point. We can explicitly find this point by extrapolating to the value of a_0 where the two components have equal amplitudes. This value of a_0 is given by

$$a_1 a_0 = \frac{3a_0^3 a_3}{4}$$

$$a_0^2 = \frac{4 a_1}{3 a_3}.\tag{4.7}$$

The IIP$_3$ is usually represented in terms of the power, so some device might have an IIP$_3$ of 10 dBm. A device with a larger IIP$_3$ has a larger linear regime.

We can also consider the output intercept point (OIP). This just adjusts the IIP by the gain of the device. Depending upon the application, one view or the other may be more useful. The OIP is larger, so it does look better on an amplifier data sheet.

As a practical matter, it is difficult to measure IIP$_3$ by watching the subtle rate of change in the amplitude of $\cos(2\pi f t)$. Other approaches are typically employed to measure IIP. For example, one can use an input of two tones and watch the growth of intermodulation (or intermod) frequencies. These new frequencies, which are produced by the nonlinearity, can be mapped back to our definition for IIP_3.

The vast majority of amplifiers used in communications are solid state, which use transistors on a semiconductor chip. Electronic amplifiers are often identified by their class. Typically, there are multiple stages of gain, but the last stage usually consumes the most power. Class A amplifiers have a final pairing of a resistor and a transistor connected to the low- and high-voltage references. This class consumes more power within the amplifier than it delivers to a load, but it is naturally fairly linear, so it has lower distortion. Class B amplifiers have transistors that both push and pull from the output signal from the low and high reference voltages. Because one or the other transistor turns on or off as needed, they can be naturally more efficient. However, it is difficult to maintain a linear input to output gain with this class. Class AB amplifiers modify class B amplifiers by adding a bias to the transistor inputs to reduce turn-on–turn-off transitions. The class D amplifier drives a low-pass filter with a switching pulse generator that modulates pulse widths or rate to drive the output signal. While these amplifiers are particularly efficient, the inductors typically employed for efficient filtering are undesirable for modern integrated circuits.

4.3   Low-Noise Amplifiers

To understand the motivation for an LNA, we need the concept of noise figure. The noise figure parameterizes the noise that is unfortunately introduced by the active circuit components in addition to the thermal noise. If there is some thermal noise power at the input of a device P_0, then the effect of the additional noise introduced by the component is given by adding $P_0 (F - 1)$ noise power to the input of an ideal component. The total effective input noise power to the ideal component is then $P_0 (F - 1) + P_0 = F P_0$. Some in the field use *noise figure* to indicate the effect on a dB scale only and use *noise factor* to indicate the effect on a linear scale. We will not use this *noise factor* convention, and we expect the reader to move between linear and dB representations as needed. As an example, if the noise figure is represented by $F = 3$ dB, we use the linear representation of $F = 2$ as needed in equations.

Once the signal gets to the receiver, everything that you do makes noise worse. The trick is to minimize the adverse effects. Let us consider a sequence or chain of amplifiers, as seen in Figure 4.3. These amplifiers could represent any sequence of devices that affect our signal. The overall gain is simply the product of the gain of the individual amplifiers. At each amplifier stage, electronics noise is contributed in addition to thermal noise.

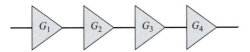

Figure 4.3 Chain of amplifiers with gains G_1, G_2, G_3, and G_4.

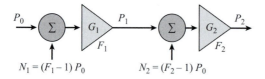

Figure 4.4 Combining noise figures for successive amplifiers.

We denote the mth amplifier's gain as G_m and the noise figure as F_m. In Figure 4.4, we depict the effect of the noise in more detail. The actual noise power at each stage is given by P_m. All powers, gain, and noise figures are expressed on a linear scale for the following evaluation. The overall noise figure of a sequence of circuit elements, which are all described as amplifiers here, is given by the ratio of the actual noise power output to the ideal thermal noise output. This overall noise figure is given by

$$F = \frac{P_n}{P_0 \prod_{m=1}^{n} G_m}. \tag{4.8}$$

By assuming that all components are essentially ideal other than the electronic noise added at each stage, we can think about what happens to the signal power and the noise. The signal power $P_{\text{sig,in}}$ is multiplied by the gain of the sequence of amplifiers, so at the output of the chain of amplifiers the signal power $P_{\text{sig,out}}$ is given by

$$P_{\text{sig,out}} = \left(\prod_{m=1}^{n} G_m \right) P_{\text{sig,in}}. \tag{4.9}$$

Similarly, if the noise figures of all amplifiers were ideal (with a 0 dB noise figure), then noise power at the output would be multiplied by the product of the gains of the amplifiers. However, each amplifier adds a bit of noise, so noise at the output is given by

$$P_n = F \left(\prod_{m=1}^{n} G_m \right) P_0, \tag{4.10}$$

where F is the combined noise figure of the sequence of amplifiers. Consequently, the output SNR is related to the ideal input SNR (where the noise is due to just thermal noise) by

$$\text{SNR}_{\text{out}} = \frac{P_{\text{sig,out}}}{P_n}$$

$$= \frac{\left(\prod_{m=1}^{n} G_m \right) P_{\text{sig,in}}}{F \left(\prod_{m=1}^{n} G_m \right) P_0}$$

$$= \frac{\text{SNR}_{\text{input}}}{F}. \tag{4.11}$$

To calculate the overall noise figure, let us first consider a single stage under the assumption that there is only noise. The noise power at the output of the amplifier is given by adding the pre-existing noise P_0 at the input to the additional component noise N_1 from the amplifier referenced at the input of the amplifier. The additional noise N_1 is related to the noise figure by

$$N_1 = (F_1 - 1) P_0. \tag{4.12}$$

If the noise figure F_1 is 0 dB (which is 1 on a linear scale), there is no additional noise: $N_1 = 0$ (linear). In general, the noise power at the output of the amplifier is given by

$$P_1 = (P_0 + N_1) G_1$$
$$= P_0 G_1 + (F_1 - 1) P_0 G_1. \tag{4.13}$$

We can solve for the overall noise figure by evaluating the ratio of the actual noise to the ideal noise, which is given by

$$F = \frac{P_1}{P_0 G_1}$$
$$= \frac{P_0 G_1 + (F_1 - 1) P_0 G_1}{P_0 G_1}$$
$$= 1 + (F_1 - 1) = F_1. \tag{4.14}$$

Unsurprisingly for a single stage, we recover an overall noise figure that is the same as the noise figure for that single amplifier.

For two amplifiers, we follow the same line of reasoning, although the process is a bit more complicated: we add noise associated with the input of the second amplifier. The slightly subtle choice is that we reference all electronic noise to the thermal noise power P_0 at the input of the first amplifier so that the contributed noise of the second amplifier is given by

$$N_2 = (F_2 - 1) P_0, \tag{4.15}$$

in terms of the noise figure of F_2. Because we want to reference additional noise at the input of any amplifier to thermal noise P_0, this is a reasonable choice for defining the noise figure. The noise at the output of the second amplifier is given by the sum of the noise power at the output of the first amplifier and the electronic noise from the second amplifier's noise figure. We can then expend these factors into terms related to the gains and noise figures of each amplifier, which is given by

$$P_2 = P_1 G_2 + N_2 G_2$$
$$= P_1 G_2 + (F_2 - 1) P_0 G_2$$
$$= [P_0 G_1 + (F_1 - 1) P_0 G_1] G_2 + (F_2 - 1) P_0 G_2. \tag{4.16}$$

The overall noise figure is given by the ratio of the total noise power at the output of the second amplifier to that given by the output of the second amplifier under an ideal assumption, which is given by

$$
\begin{aligned}
F &= \frac{P_2}{P_0\, G_1\, G_2} \\
&= \frac{[P_0\, G_1 + (F_1 - 1)\, P_0\, G_1]\, G_2 + (F_2 - 1)\, P_0\, G_2}{P_0\, G_1\, G_2} \\
&= 1 + (F_1 - 1) + (F_2 - 1)\,\frac{P_0\, G_2}{P_0\, G_1\, G_2} \\
&= F_1 + \frac{F_2 - 1}{G_1}.
\end{aligned}
\tag{4.17}
$$

We can consider N amplifiers, each with its own gain and noise figure. The noise at the output of this chain is given by

$$
P_m = (P_{m-1} + N_m)\, G_m
$$

$$
\begin{aligned}
P_N &= (P_{N-1} + N_N)\, G_N \\
&= ((P_{N-2} + N_{N-1})\, G_{N-1} + N_N)\, G_N \\
&= (((P_{N-3} + N_{N-2})\, G_{N-2} + N_{N-1})\, G_{N-1} + N_N)\, G_N \\
&= P_0 \prod_{n=1}^{N} G_n + \sum_{m=1}^{N} \left[N_m \prod_{n=m}^{N} G_n \right] \\
&= P_0 \prod_{n=1}^{N} G_n + P_0 \sum_{m=1}^{N} \left[(F_m - 1) \prod_{n=m}^{N} G_n \right].
\end{aligned}
\tag{4.18}
$$

We solve for the general noise figure by evaluating the ratio given by the Friis formula for noise figure (not to be confused with the Friis equation for propagation),

$$
\begin{aligned}
F &= \frac{P_N}{P_0\, \prod_{m=1}^{N} G_m} \\
&= \frac{P_0\, \prod_{n=1}^{N} G_n + P_0 \sum_{m=1}^{N} \left[(F_m - 1) \prod_{n=m}^{N} G_n \right]}{P_0\, \prod_{m=1}^{N} G_m} \\
&= 1 + (F_1 - 1) + \sum_{m=2}^{N} \frac{F_m - 1}{\prod_{k=1}^{m-1} G_k} \\
&= F_1 + \sum_{m=1}^{N-1} \frac{F_{m+1} - 1}{\prod_{k=1}^{m} G_k}.
\end{aligned}
\tag{4.19}
$$

One implication of the Friis formula presented in Equation (4.19) is that one should put the "best" lowest-noise-figure amplifier first to minimize the overall noise figure of a chain of amplifiers. As an example, consider two amplifiers with 10 dB gain, where amplifier A has a 3 dB (2-linear) noise figure and amplifier B has a 6 dB (4-linear) noise figure. The overall noise figure $F_{A,B}$ for A then B is given by

$$F_{A,B} = 2 + \frac{4-1}{10} = 2.3 \text{ (linear)}. \tag{4.20}$$

For comparison, the overall noise figure $F_{B,A}$ for B then A is given by

$$F_{B,A} = 4 + \frac{2-1}{10} = 4.1 \text{ (linear)}. \tag{4.21}$$

Consequently, this example demonstrates what is true in general: One should place the lowest noise-figure amplifier first to minimize the overall noise figure of a chain of amplifiers.

It is worth noting that there can be an engineering trade-off between noise figure and dynamic range for amplifiers. Dynamic range is the ratio of the largest to smallest values amplified accurately by the amplifier. Consequently, noise figure may not be the only metric for amplifier selection.

4.4 Automatic Gain Control

Given the great effort to which we go to increase our SNR, it is interesting that received signals can be too large. Specifically, the receiver has some limited dynamic range. The limits may be due to the LNA or other active elements in the radio, such as mixers or analog-to-digital converters (ADC) that saturate when the signal is too large. Ideally, the transmit power is controlled to ensure that a reasonable received signal power is observed. Transmit power control does require a feedback signal to be sent to the transmitter. This communications channel does not exist for all systems.

Conversely, the receiver can adaptively attenuate the received signal. This has the effect of reducing the SNR, which is typically considered undesirable. However, compared to a saturated signal, an attenuated signal is often preferable. Traditionally, this automatic gain control (AGC) is performed with an analog circuit, although modern systems now often use a digitally controlled gain or attenuation to perform this function.

Problems

4.1 Express the following gains and noise figures on a linear scale:
(a) $G = 10$ dB
(b) $F = 6$ dB
(c) $G = 23$ dB
(d) $F = 3$ dB
(e) $G = 17$ dB

4.2 Given the sequential chain of ideal amplifiers with gains of 10 dB, 10 dB, and 6 dB, evaluate:

(a) the overall gain of the chain of amplifiers; and

(b) the output voltage of the chain of amplifiers if the input voltage is 1 mV.

4.3 For a power amplifier that operates to the $\pm V$ supply voltages, if the supply voltages are ± 5 V and the input signal has voltages bounded by $\|s(t)\| < 100$ mV, what is the largest possible amplifier gain?

4.4 Consider a signal with output powers of 10^{-6} W, 0.4 W, 10 W, and 20 kW. Express these powers in terms of dBW and dBm.

4.5 If the limits of a sine wave are ± 0.1 V at the input of an amplifier and ± 2 V at its output, what is the amplifier gain both on linear and decibel scales?

4.6 If the power ratio of signal to thermal noise at the input of an amplifier is 10 dB, what are the SNRs at the output of the amplifier for noise figures of 0 dB, 2 dB, and 6 dB?

4.7 By considering a chain of three amplifiers with gains G_1, G_2, and G_3 with noise figures F_1, F_2, and F_3, evaluate explicitly the overall noise figure of the systems.

4.8 Consider the chain of three amplifiers labeled A, B, and C with gains 10 dB, 13 dB, and 10 dB, with corresponding noise figures 6 dB, 3 dB, and 3 dB, respectively.

(a) What is the best order of amplifiers (why)?

(b) What is the overall gain of the chain of amplifiers?

(c) What is the best overall noise figure (show your calculation)?

4.9 For amplifiers A, B, C, and D with gains 6 dB, 10 dB, 10 dB, and 6 dB, and with corresponding noise figures of 3 dB, 3 dB, 6 dB, and 6 dB:

(a) What is the overall gain?

(b) What is the best ordering? Why?

(c) What is the corresponding overall noise figure?

4.10 What is the noise figure of an infinite sequence of amplifiers if each amplifier has a gain of 10 dB and a noise figure of 3 dB?

4.11 Consider the simple line-of-sight radio link between antennas pointed at each other with the following operating parameters:

• transmit power: 10 mW

• transmit gain: 2 dBi

• receive gain: 13 dBi

• carrier wavelength: 1 m

• range: 40 km

• bandwidth: 20 MHz

• noise figure: 6 dB

Assume the receiver is at room temperature (~ 300 K).

(a) Evaluate the link distance in number of wavelengths, expressed in dB.

(b) Evaluate the attenuation in power from the transmitter to the receiver.

(c) Evaluate the received signal power.

(d) Evaluate the SNR in the receiver, including noise figure.

(e) Evaluate the best-case data rate in Mb/s.

4.12 By using a numerical simulation, evaluate the noise power of a chain of three amplifiers with gains of 20 dB, 10 dB, and 10 dB, each with noise figures of 6 dB. Assume that there is only complex Gaussian noise (that is, no signal) at the input of the chain. Use units such that the thermal input noise has a variance of 1.

5 Up- and Down-Conversion

In this chapter, we discuss the concepts of baseband and passband representations of signals and the mechanisms for moving between these two forms by using up- and down-conversion. We describe multiple up- and down-conversion approaches, such as digital-only, direct, superheterodyne, and digital intermediate frequency (IF). We discuss the components used to move between baseband and passband representation: analog-to-digital and digital-to-analog converters (ADC and DAC, respectively), and frequency synthesizers.

5.1 Meaning of Baseband and Passband

The concepts of baseband, passband, and the up- and down-conversion that relates the two bands are critical to understanding wireless communications systems.

5.1.1 Baseband

We typically think of our baseband signal as being close to the information that we wish to send. In Chapter 6, we will discuss modulation approaches to convert bits to complex baseband signals. The spectrum of the baseband signal is often centered around DC[1] (0 Hz in frequency), with both positive and negative frequency contributions. This signal is complex in general; thus, we often employ the phrase *complex baseband* to describe it. The motivation for using complex numbers is related to the observation that sine and cosine signals are orthogonal. Consequently, we use complex numbers to associate real signals with the sine and cosine components used by up-conversion, as is motivated by Euler's formula. The technology that converts from baseband to passband or passband to baseband is sometimes called *heterodyne* or *superheterodyne modulation*. There is some ambiguity in the use of the term *modulation*. Historically, for analog communications systems, the encoding of the information and the up-conversion were performed in a single stage; consequently, *modulation* was often used to indicate what we call up-conversion. In an attempt to

---

[1] Direct current (DC) is used to indicate the static or zero-frequency component of a signal. Its etymology comes from a description of power sources, but the initialism DC is used broadly.

be more precise, we will use the word *modulation* to indicate mapping bits to complex baseband, and the phrases *up-* and *down-conversion* to indicate moving between baseband and passband frequencies.

5.1.2   Passband

There are multiple reasons to move communications signals from complex baseband to some passband frequency. We depict the typical transmitter and receiver radio block diagrams in Figure 5.1. First, wireless systems cannot transmit signals at 0 Hz frequency (DC). Second, while there are some subtleties in this statement, lower frequencies typically require larger antennas to be efficient. If one wishes to employ a handheld communications system, having a 10 m antenna is probably undesirable. Finally, various frequency bands are allocated to different users. In the United States, these spectral allocations are regulated by the Federal Communications Commission (FCC).

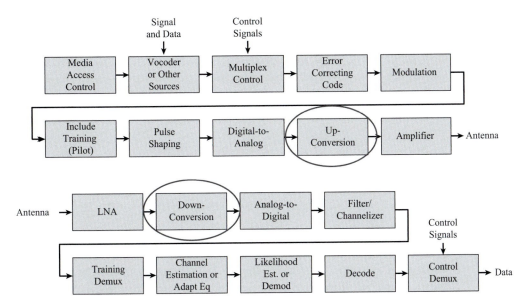

Figure 5.1 Up- and down-conversion blocks in standard radio transmit and receive processing chains.

We generally convert the signal that was generated at baseband with some bandwidth to passband, which is centered at some carrier frequency with some bandwidth. We review moving the spectrum of signals in the frequency domain in Section 15.4.2, and depict it in Figure 15.3. One issue with the discussion in Section 15.4.2 is that the signal after this conversion is complex in general. While complex signals are useful to consider in processing and in theory, the passband signal transmitted or received must be real to be physically transmitted.

5.2   Up- and Down-Conversion

5.2.1   Up-Conversion

If we have a time-domain complex baseband signal $s(t)$ with corresponding frequency-domain representation $S(f)$, then the time-domain version of a frequency shifted signal is given by

$$\mathcal{F}^{-1}\{S(f - f_c)\} = e^{i\,2\pi f_c\,t}\,s(t)\,. \tag{5.1}$$

It is worth emphasizing that this signal is complex in general because both $s(t)$ and the exponential are complex functions. As we noted previously, we cannot transmit or receive complex signals through antennas, so we must convert this to a real signal.

Fortunately, the real component of the complex frequency shifted signal contains all the information necessary to reproduce the complex baseband signal at the receiver. To help us track the real and imaginary components of $s(t)$, let us use the notation

$$s(t) = x(t) + i\,y(t)\,, \tag{5.2}$$

where $x(t)$ and $y(t)$ are real functions. These signals are sometimes respectively called the *in-phase* and *quadrature* components of the communications signal at baseband. Consequently, the complex baseband representation is sometimes called an *I–Q representation*. If we denote the time-domain signal at passband $s_{pb}(t)$, then the real part of the frequency shifted signal is given by

$$
\begin{aligned}
s_{pb}(t) &= \Re\{\mathcal{F}^{-1}\{S(f - f_c)\}\} \\
&= \Re\{e^{i\,2\pi f_c\,t}\,s(t)\} \\
&= \Re\{[\cos(2\pi f_c\,t) + i\,\sin(2\pi f_c\,t)]\,[x(t) + i\,y(t)]\} \\
&= \cos(2\pi f_c\,t)\,x(t) - \sin(2\pi f_c\,t)\,y(t)\,.
\end{aligned} \tag{5.3}
$$

As we observed previously, the sine and cosine functions are orthogonal, so we can, in theory, recover the information contained in both $x(t)$ and $y(t)$. Interestingly, the use of $e^{i\,2\pi f_c\,t}$ versus $e^{-i\,2\pi f_c\,t}$ in Equation (5.4) is arbitrary, so long as the transmitter and receiver agree. If we were to use the alternative convention, the passband signal would be given by

$$
\begin{aligned}
s_{pb}(t) &= \Re\{\mathcal{F}^{-1}\{S(f + f_c)\}\} \\
&= \Re\{e^{-i\,2\pi f_c\,t}\,s(t)\} \\
&= \cos(2\pi f_c\,t)\,x(t) + \sin(2\pi f_c\,t)\,y(t)\,.
\end{aligned} \tag{5.4}
$$

For the sake of our discussion and analysis, we keep with the former convention of $\Re\{e^{i\,2\pi f_c\,t}\,s(t)\}$, but it is important to remember that either might be employed in any given system.[2]

---

[2] I actually once worked on a system that used one convention below 3 GHz and flipped above 3 GHz.

Equivalently, we can construct the real component of the frequency shifted signal by adding the signal with the complex conjugate of the signal and scaling by a factor of 1/2. The passband signal is then given by

$$s_{pb}(t) = \Re\{e^{i2\pi f_c t} s(t)\}$$

$$= \frac{e^{i2\pi f_c t} s(t) + e^{-i2\pi f_c t} s^*(t)}{2}.   \tag{5.5}$$

Consequently, the spectrum of the passband signal $S_{pb}(f)$ is given by

$$S_{pb}(f) = \mathcal{F}\{s_{pb}(t)\}$$

$$= \frac{\mathcal{F}\{e^{i2\pi f_c t} s(t)\} + \mathcal{F}\{e^{-i2\pi f_c t} s^*(t)\}}{2}$$

$$= \frac{S(f - f_c) + \int dt\, e^{-i2\pi [f + f_c] t} s^*(t)}{2}$$

$$= \frac{S(f - f_c) + \left[\int dt\, e^{-i2\pi [-f - f_c] t} s(t)\right]^*}{2}$$

$$= \frac{S(f - f_c) + \left[\int dt\, e^{-i2\pi [f'] t} s(t)\right]^*}{2} ;\quad f' = -f - f_c$$

$$= \frac{S(f - f_c) + [S(f')]^*}{2}$$

$$= \frac{S(f - f_c) + S^*(-f - f_c)}{2}.   \tag{5.6}$$

In the spectral domain, this process produces signals of the bandwidth of $s(t)$ centered at both $+f_c$ and $-f_c$, as is depicted in Figure 5.2.

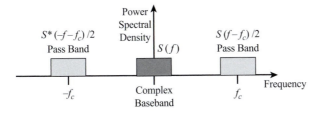

Figure 5.2 Passband spectrum after up-conversion processing.

5.2.2   Down-Conversion

If we denote the received signal at passband by $z_{pb}(t)$, then we can recover a distorted version of the baseband signal by multiplying (or mixing) the received signal and sinusoids at the carrier frequency. This mixing causes an undesirable spectral image at around twice the carrier frequency. We can remove this image by applying a low-pass filter, as discussed in Section 2.2.5. We recover the complex baseband version of the received signal $z(t)$ by using

$$z(t) = \mathcal{LPF}\{z_{\text{pb}}(t) \left[\cos(2\pi f_c t) - i\,\sin(2\pi f_c t)\right]\}, \tag{5.7}$$

where we represent the low-pass filter with $\mathcal{LPF}\{\cdot\}$.

As a hypothetical example, consider receiving a pure tone in the absence of any noise that is place at DC in baseband, so that the received signal is given by

$$z_{\text{pb}}(t) = a_c\,\cos(2\pi f_c t) - a_s\,\sin(2\pi f_c t), \tag{5.8}$$

where a_c and a_s are the real coefficients associated with the received cosine and sine terms, respectively. By multiplying the hypothetical received sinusoidal signal with a sinusoid at the carrier frequency, the mixed signal is given by

$$\begin{aligned}
\cos(2\pi f_c t)\,z_{\text{pb}}(t) &= \cos(2\pi f_c t)\left[a_c\,\cos(2\pi f_c t) - a_s\,\sin(2\pi f_c t)\right] \\
&= a_c\,\cos^2(2\pi f_c t) - a_s\,\sin(2\pi f_c t)\,\cos(2\pi f_c t) \\
&= a_c\,\frac{1 + \cos(2\pi\,2f_c t)}{2} - a_s\,\frac{\sin(2\pi\,2f_c t)}{2}.
\end{aligned} \tag{5.9}$$

The terms at twice the carrier frequency are not of interest to us. We can remove them by applying a low-pass filter with a corner frequency well below twice the carrier frequency. The filtered version of the mixed receive signal is given by

$$\begin{aligned}
\mathcal{LPF}\{\cos(2\pi f_c t)\,z_{\text{pb}}(t)\} &= \mathcal{LPF}\left\{a_c\,\frac{1 + \cos(2\,2\pi f_c t)}{2} - a_s\,\frac{\sin(2\,2\pi f_c t)}{2}\right\} \\
&= \frac{a_c}{2}.
\end{aligned} \tag{5.10}$$

Similarly, if we consider mixing the received signal with the sine, then a_s is recovered, as seen by

$$\begin{aligned}
\mathcal{LPF}\{-i\,\sin(2\pi f_c t)\,z_{\text{pb}}(t)\} &= -i\,\mathcal{LPF}\left\{a_c\,\frac{\sin(2\,2\pi f_c t)}{2} - a_s\,\frac{1 - \cos(2\,2\pi f_c t)}{2}\right\} \\
&= i\,\frac{a_s}{2}.
\end{aligned} \tag{5.11}$$

For convenience, we can associate a_c with the real component and a_s with the imaginary component. With this definition, we have constructed our complex baseband signal, given by

$$z(t) = \frac{a_c + i\,a_s}{2}. \tag{5.12}$$

Imagine that the parameters a_c and a_s are varying slowly compared to the sinusoid associated with the carrier frequency. We can then recover data that is encoded in the received signal.

5.3   Digital-Only Up- and Down-Conversion

This mathematical form for up-conversion processing would have little value if we could not implement it. We can realize this form by digitally up- and down-converting and then converting between digital and analog radio frequencies (RFs), as we depict in Figures 5.3 and 5.4. This approach requires very high-frequency digital-to-analog converters (DACs) and analog-to-digital converters (ADCs) that operate at a rate that supports RF directly. Regularly sampling DACs and ADCs have to operate at a frequency of greater than $2f_c + B$ for a carrier frequency of f_c and a baseband signal bandwidth of B. As the performance of DACs and ADCs improves, the RFs that can be supported by directly sampling the passband signal continue to increase. However, it is typically more power efficient to use other approaches.

Transmission

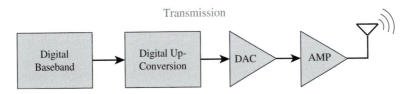

Figure 5.3 Digital-only up-conversion.

Reception

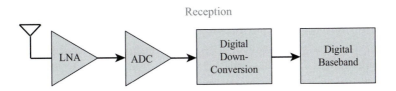

Figure 5.4 Digital-only down-conversion.

5.4   Direct Up- and Down-Conversion

We can also realize this up- and down-conversion by using a direct-conversion circuitry. In Figure 5.5, we present the direct up-conversion block diagram. It splits the real and imaginary components of the baseband signal into two physical chains. We use a given modulation approach in the digital processor to convert the information bits into a digital complex baseband signal. The real and imaginary components are separately converted from digital signals to analog signals (described in Section 5.7). The real and imaginary analog signals are multiplied (or, equivalently, mixed) by cosine and sine waveforms at the carrier frequency. The sine is constructed from the cosine by employing a 90-degree phase shift. After mixing the signals associated with the real and imaginary chain, the results are summed to produce a passband signal. This signal is then amplified and transmitted.

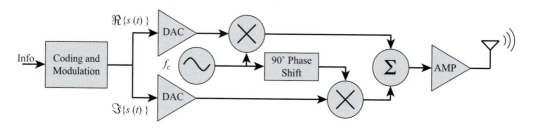

Figure 5.5 Direct up-conversion processing.

One of the advantages of direct conversion (as opposed to approaches that employ superheterodyne up-conversion) is that the filtering that might be employed is typically not particularly aggressive. Consequently, the conversion can be employed on the same integrated circuit as the rest of the RF circuitry, which is of significant potential value in terms of radio size, weight, and power consumption. One of the potential disadvantages is that the real and imaginary chains are never exactly the same. As a result, the transmitted signal will be distorted by I–Q mismatch.

Unsurprisingly, the block diagram for direct down-conversion processing is similar to that for up-conversion, but backwards (Figure 5.6). For some applications, there is some pre-selection filter after the receive antenna that reduces the effects of strong out-of-band signals that might overwhelm the first active component, which is typically a low-noise amplifier (LNA), as discussed in Section 4.3. The received signal is then split into two channels. One channel is mixed with a cosine and the other is mixed with a sine waveform. To remove the high-frequency contributions, we apply a low-pass filter to both the real and imaginary components. The signals are then converted to a digital signal using ADCs. The digitized signal is then processed to recover the transmitted information.

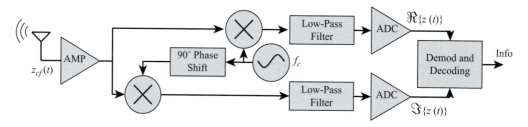

Figure 5.6 Direct down-conversion processing.

5.5   Superheterodyne Conversion

While direct conversion is often selected for physically small applications, there are advantages to employing more complicated conversion approaches. In Figures 5.7

and 5.8, we depict superheterodyne down- and up-conversion, respectively. The superheterodyne converter adds a stage to the conversion. The first level of conversion is between baseband and an intermediate frequency (IF). The next level of conversion is between IF and the passband. Some systems implement multiple IF conversions. There are multiple potential advantages to this approach. One can potentially design a very high-quality analog IF filter that mitigates undesired out-of-band signals. The carrier frequency used can be easily changed by changing the mixing frequency. This change can be made with direct conversion as well; however, this form allows for the reuse of the same analog IF filter even as the carrier frequency is changed. For the receiver, the conversion between the passband frequency and IF is followed by a direct conversion receiver that takes the IF and converts it to complex baseband. To be clear, that is not the version that is described traditionally. For analog communications systems, the IF frequency is fed into another analog down-conversion stage that incorporates an analog demodulator. The analog amplitude modulation (AM) and frequency modulation (FM) used for older broadcast radio systems use these approaches. For the sake of brevity, we leave that discussion to other texts.

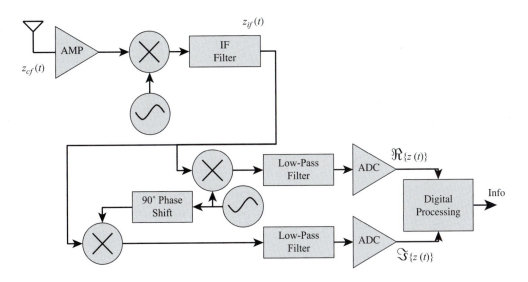

Figure 5.7 Superheterodyne down-conversion processing.

One of the issues to consider with a superheterodyne system is frequency planning. Because mixers are typically nonlinear, the mixing of signals often produces harmonics. Consider a received tone at f_c and a mixed tone at f_1. The output of the mixer potentially has all contributions at all frequencies $mf_c \pm nf_1$ for all positive integers m and n. If the IF filter is perfect, these harmonics are unlikely to be an issue. However, no filter is perfect, and some energy is bound to leak past it. At the second

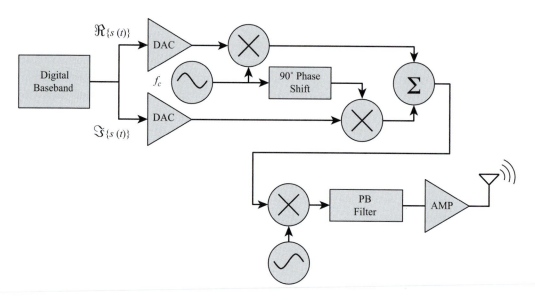

Figure 5.8 Superheterodyne up-conversion processing.

mixer, these leaked harmonics can potentially mix with harmonics of the second mixer to contaminate the baseband signal. Consequently, the two frequencies should be selected so that the lower mixed harmonics do not fall into the baseband. If the second mixer uses f_2, then frequencies $(m f_c \pm n f_1) k \pm j f_2$ for all positive integers k and j. Typically, the system is designed under the assumption that $m = n = k = j = 1$, so that $f_c \pm f_1 \pm f_2 = 0$. While either the sum or difference can be used for both the f_1 and f_2 terms, the most common approach is to set the frequencies so that $f_c - f_1 - f_2 = 0$. For those undesired harmonics that significantly contribute to mixer output, we want

$$\|(m f_c \pm n f_1) k \pm j f_2\| > B/2; \quad m, n, k, j \in \mathbb{Z}, \tag{5.13}$$

where B is the bandwidth of the baseband filter.

5.6   Superheterodyne Conversion with Digital IF

In modern high-performance systems, superheterodyne implementations with digital IF are not uncommon, as depicted in Figures 5.9 and 5.10. This implementation enables the removal of separate in-phase and quadrature DACs or ADCs. The IF signal is captured or constructed digitally. Consequently, the conversion can be performed with high fidelity, and potential concerns about I–Q imbalance are effectively removed. However, this implementation requires higher sampling rates for the DACs and ADCs.

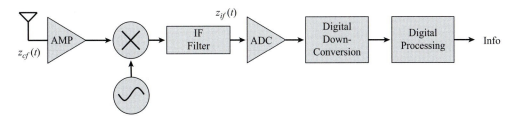

Figure 5.9 Superheterodyne down-conversion processing with digital IF.

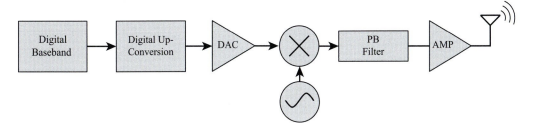

Figure 5.10 Superheterodyne up-conversion processing with digital IF.

5.7   Analog to/from Digital Conversion

Conceptually, DACs and ADCs are relatively simple. For a DAC, a digital value is converted to a corresponding voltage or current. Typically, the output is updated in time at a regular interval. To compensate for the switching between discrete levels, an output low-pass filter is typically employed to remove out-of-frequency-band artifacts. Similarly for an ADC, a real band-limited analog voltage is assigned a digital value that most closely matches the observed analog voltage. Because ADCs cannot operate instantaneously, they often employ a sample-and-hold or a track-and-hold circuit prior to the actual ADC. This circuit holds the value for a duration sufficient for the ADC to accurately convert the voltage to a digital representation. These circuits are often integrated into the ADC package, so their presence might not be clear.

Real ADCs and DACs have limited sample rates and dynamic ranges. Here, dynamic range indicates the ratio of the largest to smallest change in values that the converter can accurately represent, as is indicated in Figure 5.11. If a converter has n_b bits to represent a number, then it can encode 2^{n_b} levels. As an example, consider a 14-bit converter with equally spaced voltage levels. While the exact number is dependent upon the signal that is being represented, the dynamic range (DR) in power is given by

$$DR = \left(\frac{2^{n_b}}{1}\right)^2 \tag{5.14}$$

$$\approx n_b \cdot 6 \text{ dB on a dB scale.} \tag{5.15}$$

(Here we really are mixing dB and linear calculations.) The dynamic range doubles for each additional bit; consequently, the 14-bit converter should have approximately 84 dB of dynamic range, which is sufficient for most communications applications. However, the reality is that most converters are not accurate to the number of bits that can be specified, so there is often an effective number of bits that is used to indicate the converter's performance. For example, a 14-bit ADC may only have 12 effective bits because the smallest couple of bits are not sufficiently accurate to be all that useful. Furthermore, for some given state of technology and power consumption, higher sample rate converters typically have a lower number of effective bits. The communications engineer must select the converter characteristics based upon multiple metrics of performance.

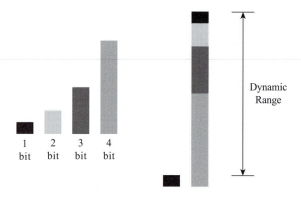

Figure 5.11 Dynamic range of digitized signal.

5.7.1   Setting the Noise Level

The finite dynamic range of the ADC creates another engineering constraint. If we want to make full use of this dynamic range, we need to adjust the gains prior to the ADC in the receive chain so that the noise is at the correct level. In the best-case scenario, we set the variance of the noise by adjusting this gain so that the noise causes the couple of bits associated with the smallest levels to fluctuate, but not more than that. Variations in circuit temperature and aging can make it difficult to maintain this goal unless the system is automatically self-calibrating.

5.8   Frequency Synthesizer

We have assumed that we had access to arbitrary local oscillator frequencies. It is easy to write mathematically that we have access to an oscillator with frequency f_c, but it is more complicated in practice. Frequency offsets and phase uncertainties produced by the synthesizer often drive the waveform design and place practical limits on communications performance. While there are multiple types of frequency synthesizers, we consider a slightly simplified version of a common approach. We use a phase-lock loop (PLL) to transform a reference, typically produced with a crystal oscillator at a lower frequency, to some rational ratio times the crystal frequency. As an example, we consider an oscillator that produces a reference of 10 MHz. If we want a carrier frequency of 1795 MHz, we can get to this frequency by multiplying the reference frequency by the ratio of 3590/2.

Crystal oscillators are not perfect devices. Their frequencies are sensitive to temperature and, to some extent, age. They also do not produce a pure frequency, and generate phase noise. The phase, and consequently the frequency, wanders over time. The effect of phase noise is often expressed in terms of the power spectral density relative to the power of the intended frequency (expressed in units of dBc/Hz, which is a slightly confusing notation) as a function of frequency offset. The unit dBc indicates power in dB relative to the carrier frequency. Low-frequency phase noise can limit how long we can assume an estimated channel phase is valid. High-frequency phase noise limits the maximum order of the constellation, which we discuss in Section 6.1. The high-frequency phase noise causes effective rotations of the constellation from chip to chip.

There are multiple types of crystal oscillators and numerous standard frequencies. The oscillators are often produced and then trimmed or sorted to achieve a given frequency accuracy. Example oscillators include a simple crystal oscillator (XO), a temperature-compensated crystal oscillator (TCXO), and an oven-controlled crystal oscillator (OCXO). We often discuss the performance of oscillators in terms of absolute error and in terms of their temperature stability. Simple crystal oscillators have relative errors that are on the order of several parts in 10^6. As an example, consider an error of 3×10^{-6}. If we have a carrier frequency of 1 GHz, then we would expect a 3 kHz frequency offset. Temperature-compensated crystal oscillators often have performance that is on the order of less than one part in 10^6. They have some "intelligence" that attempts to compensate for the errors induced by the temperature changes. While most communications systems are not affected by it, the compensation itself can cause problems for particularly sensitive applications. As one might expect, OCXOs have a small thermally isolated container with temperature-controlled heating elements that attempt to keep the temperature within a narrow range. They may have fixed frequency offset in the area of one part in 10^6, but stability over time that achieves one part in 10^8 or better. Finally, while not common at the time of this writing, chip-scale atomic clocks are an active area of research, and they have excellent absolute frequency accuracy. Depending upon the application, we can select the appropriate oscillator.

With a given frequency reference, we use a PLL to synthesize our carrier frequency. The PLL seems nearly magical. In Figure 5.12, we depict the block diagram of a simple synthesizer. The reference frequency, which is typically produced by a crystal oscillator, is divided by a factor of M by the use of a counter. As an example, the counter cycles the output voltage every M input cycles. The output of this counter is mixed (multiplied) by the output of a counter that is driven by the synthesizer's output signal. The counter divides the output frequency by N. Thus, if the synthesizer is locked, then the mixer multiplies a signal at f_{XO}/M and f_c/N. When locked, these two derived frequencies are equal, such that

$$f_{XO}/M = f_c/N$$

$$f_c = \frac{N}{M} f_{XO}. \qquad (5.16)$$

The magic is achieved by observing that the output of the mixer has the sum and difference of the input frequencies. A loop filter removes the sum frequencies, leaving the near-DC component. This low-frequency component drives the input of a voltage-controlled oscillator (VCO), which maps an input voltage to an output frequency. Prior to lock, the energy spectral density will fluctuate significantly, but will quickly settle to the correct frequency. This settling time must be incorporated into the design of a system that changes carrier frequencies.

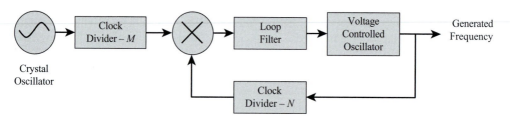

Figure 5.12 Block diagram of a frequency synthesizer that produces a frequency that is N/M the crystal oscillator reference frequency.

Problems

5.1 By using direct conversion with a carrier frequency f_c, up-convert the complex baseband signal given by the low-frequency complex tone $e^{i\,2\pi\,f_{tone}\,t}$.
(a) Draw the up-conversion block diagram.
(b) Express the time-domain signal at passband.
(c) Draw the spectrum at passband (both positive and negative frequencies).

5.2 A frequency-domain passband signal is represented by

$$\frac{S^*(-f - f_c) + S(f - f_c)}{2}.$$

(a) Draw the direct down-conversion block diagram.
(b) Draw the shape of the power spectrum density before the application of the low-pass filters (both positive and negative frequencies) if the power spectral density of $S(f)$ can be represented approximately by a top hat shape.
(c) Show that output of down-conversion is proportional to $S(f)$ if the bandwidth of $S(f)$ is small compared to f_c.

5.3 What is the approximate number of effective bits of an ADC if the ratio of the largest to smallest power differences that can be accurately represented is 10^6?

5.4 For a communications system that operates at a carrier frequency of 1 GHz and bandwidth of 10 MHz, what are the minimum DAC and ADC sample rates for:
(a) a digital-only (no mixers) conversion approach; and
(b) a direct-conversion approach.

5.5 Consider the up-conversion of a DC baseband signal of unit amplitude to normalized frequency $f = 0.3$.
(a) In simulation, digitally up-convert 1000 pseudorandom samples of a DC signal and plot an estimate of the power spectral density for the baseband signal.
(b) In simulation, digitally up-convert 1000 samples of a DC signal and plot an estimate of the power spectral density for the passband signal.

5.6 For a line-of-sight link, evaluate the SNR for a 100 mW transmitter with transmit and receive antennas with 0 dBi. Assume that the link length is 1 km, the bandwidth is 20 MHz, the carrier frequency is 2.4 GHz, the temperature is 300 K, and the noise figure is 6 dB.

5.7 Given some complex baseband signal $s(t)$ with corresponding baseband frequency-domain representation $S(f)$ and a carrier frequency f_c, derive the passband spectrum in terms of $S(f)$ and f_c.

5.8 Consider a digital superheterodyne receiver. Develop a frequency plan for a carrier frequency of 10 GHz and a baseband filter bandwidth of 10 MHz, under the assumption that the fundamental, second, and third harmonics contribute.

5.9 Consider frequency plans for a digital superheterodyne receiver. If the received passband signal is proportional to $\Re\{s(t)\,e^{i\,2\pi f_c t}\}$, that is mixed with a tone at f_1 to move the carrier frequency to the IF frequency f_{IF}, under what conditions is the recovered baseband signal proportional to $s(t)$ versus $s^*(t)$?

5.10 For a frequency synthesizer generated from a 40 MHz reference, specify the input divider and loop divider values for the following carrier frequencies:
(a) $f_c = 140$ MHz
(b) $f_c = 1710$ MHz
(c) $f_c = 2415$ MHz

5.11 Consider a frequency synthesizer that is generating a carrier frequency of 10 GHz from a 10 MHz crystal reference. The baseband signal is a DC signal of 1 and the error of the crystal frequency is 10^{-6} of the reference frequency.
(a) What is the error at the carrier frequency?
(b) At the receiver at complex baseband, how long does it take for the transmitted signal to reverse sign compared to the intended phase?

5.12 Given currently commercially available parts, design an up- and down-conversion approach for a communications system operating at a carrier frequency of 28 GHz with 100 MHz bandwidth that is using a modulation requiring at least 36 dB of dynamic range. Specify the approach and the specific parts used.

6   Modulation and Demodulation

In this chapter, we consider how to map bits to a sequence of voltages or signal levels at complex baseband. We introduce the idea of a constellation, which defines a lattice of allowed baseband signaling voltages. We provide a discussion of modulation-specific capacities. We introduce the idea of pulse shaping that we use to reduce the spectral spread of our signals. We discuss channel estimation and compensation. We evaluate the raw bit error rate of BPSK and symbol error rates of QPSK modulations. We discuss demodulation and consider both hard decisions and soft decisions, which involves estimating likelihoods of possible symbols.

In this chapter, we focus on single-carrier modulation under the assumption of a flat-fading channel, so the model is given by

$$z(t) = a\,s(t) + n(t)\,, \tag{6.1}$$

where $z(t)$ is the complex baseband received signal and $n(t)$ is additive Gaussian noise. In Chapter 7, we consider more complicated channels. Unlike in our discussion in Chapters 2 and 3, here we assume the system gains are incorporated into $\|a\|^2$ and that $E\{\|s(t)\|^2\} = 1$. It is often useful to use units of power such that we have unit-variance noise $E\{\|n(t)\|^2\} = 1$.

6.1   Constellations

Modulation is the mapping between bits and voltages. As indicated in Figure 6.1, We modulate in the transmitter and demodulate in the receiver. In modern systems, a decision is not made as each symbol is received, but rather a likelihood of each possible modulated symbol is estimated. This is often termed *hard* versus *soft demodulation* or *hard* versus *soft decisions*, which will be discussed further in Chapter 8.

When we constructed the bound for the channel capacity, discussed in Section 3.1, we assumed that signals were drawn from a Gaussian distribution. While systems can work by using this approach, practically it is often easier to work with a signal that is drawn from a discrete set (or lattice) of levels. By thinking about the complex baseband signal, we enable a set of levels for both the real and imaginary components. These values are updated at the symbol rate or the chip rate. Because

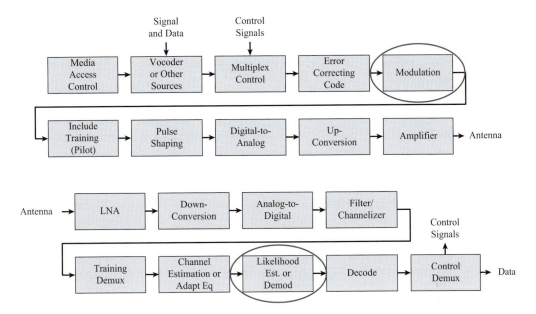

Figure 6.1 Representation of a typical communication system block diagram with modulation and demodulation blocks indicated.

we can mean different things by using the term *symbol*, the use of chip rate is less ambiguous. We call the set of allowed values the constellation.

6.1.1   Gaussian Signal

Let us first consider a Gaussian signal. Imagine that we have a sequence of n_{chip} values transmitted regularly at the chip rate. We build a dictionary of n_{seq} sequences drawn from a Gaussian distribution. The transmitted signal is attenuated by the propagation channel and noise is added at the receiver. The receiver can test the observed received sequence by determining the distance from it and all of the sequences in the dictionary. The decoder identifies the one that is closest by some distance metric as the correct one, if we assume that each sequence is equally likely. As we previously discovered, we can decode without errors as the number of chips approaches infinity if the size of the dictionary is below some threshold that is dependent upon the signal-to-noise ratio (SNR). This is the approach that we discussed to motivate the Shannon limit in Section 3.2. The number of bits n_{bits} of information conveyed in this approach is given by

$$n_{bits} = \log_2(n_{seq}),\qquad(6.2)$$

which is the number of bits required to index a given sequence. Consequently, the data rate per chip – or, equivalently, per symbol or per channel usage – r (bit/chip) is given by

$$r = \frac{\log_2(n_{\text{seq}})}{n_{\text{chip}}}. \qquad (6.3)$$

The actual bit rate R is given by the product of the bits per chip r (b/chip) and the rate of chips R_{chip} (chip/s), so that

$$R = r R_{\text{chip}} \text{ (b/s)}. \qquad (6.4)$$

It is theoretically possible to fit a sequence of independent symbols produced at a rate of R_{chip} in a double-sided bandwidth of $B = R_{\text{chip}}$, although in practice you always need a bit more bandwidth. One spectral confinement approach is to use pulse shaping to constrain the bandwidth employed. We discuss this idea in greater detail in Section 6.2. If we optimistically assume that bandwidth is given by the chip rate, then Equation (6.3) is the actual spectral efficiency of the communications system. This discussion is also valid if each symbol or chip is limited to a discrete set of values. The number of discrete options n_{opt} for each chip places an upper limit on the possible rate $r \leq \log_2(n_{\text{opt}})$; however, for practical systems, this choice of a finite set of modulation options can simplify processing.

It is easy to become confused about the difference between the number of possible options for a given symbol (which is typically relatively small) and the number of possible sequences of symbols that includes all possible allowed combinations of symbol sequences over some number of chips (which is typically quite large). The latter is often used in proofs and then normalized by the number of chips to get an average number of options per symbol, as described in Equation (6.3).

6.1.2   Constellation Diagram

In Figure 6.2, we present examples of discrete modulation approaches. In the up-conversion process, the cosine and sine components of the carrier sinusoid are modified by the real and imaginary components of symbol options. These values are proportional to the complex baseband in-phase and quadrature modulating voltages used by the up-conversion block that we discussed in Chapter 5. The points on these diagrams indicate the set of allowed symbol locations. For any given symbol, only one of these points is employed. The set of allowed points is often termed a *constellation*. The type of constellation is often denoted "something" keying. The word *keying* is a historical reference to the old use of Morse code, tapped by hand on a switch called a key.

In Figure 6.2, we display binary-phase-shift keying (BPSK), quadrature-phase-shift keying (QPSK), 8-phase-shift keying (8-PSK), and 16-quadrature-amplitude modulation (16-QAM). Each constellation has n_{point} points. We indicate the complex amplitude of the constellation point as m_n for the nth constellation point, so $n \in \{1, \ldots, n_{\text{point}}\}$. In general, you could lay down any pattern you wished and create a new constellation; however, these are the common modulation schemes. The number of bits that each constellation could possibly send is given by $\log_2(n_{\text{point}})$.

Consequently, the maximum number of bits sent for each one of these constellations is 1, 2, 3, and 4 for BPSK, QPSK, 8-PSK, and 16-QAM, respectively. As a simple example, we map a two-bit sequence to one of four QPSK symbols. We might label

$$(0,0) \Rightarrow m_1 = \frac{1+i}{\sqrt{2}} \tag{6.5}$$

$$(0,1) \Rightarrow m_2 = \frac{-1+i}{\sqrt{2}} \tag{6.6}$$

$$(1,0) \Rightarrow m_3 = \frac{-1-i}{\sqrt{2}} \tag{6.7}$$

$$(1,1) \Rightarrow m_4 = \frac{1-i}{\sqrt{2}}. \tag{6.8}$$

For a regularly sampled signal with separation T, so that $t = kT$ for the kth sample, our complex baseband signal $s(t)$ at the kth sample is given by

$$s(kT) = m_{n(k)}, \tag{6.9}$$

where we use $n(k)$ to indicate the constellation index n at the kth sample in time.

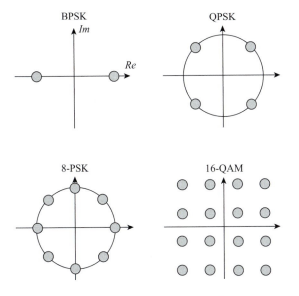

Figure 6.2 Representation of BPSK, QPSK, 8-PSK, and 16-QAM constellations.

Constellations with a larger number of options are typically more difficult to decode than those with a smaller number. For the sake of discussion, we assume that each constellation is scaled to have unit variance (same average power). It is then easy to scale the signal to provide the actual transmit power. Imagine that each point in the constellation has some fuzzy circle of noise around the point. As the number of points in a constellation increases, the distance between points is reduced, and the noise is more likely to cause confusion between points. Error correcting codes incorporate constraints so that not all sequences of constellation points are allowed. This has the effect of reducing both the data rate and the error rate.

A second point is that we can potentially send more bits than one would naively think if we are operating in a very high SNR environment. The 16-QAM constellation enables up to 4 bits per chip (or, equivalently, per channel usage). If we do a perfect job constraining the bandwidth of the signal with ideal pulse shaping, which is discussed in Section 6.2, then we have one complex channel usage at rate R_{chip} that only requires bandwidth $B = R_{chip}$. Consequently, the limiting spectral efficiency is 4 b/chip $= 4b/(s \cdot Hz) = 4b/s/Hz$ and ideal spectrum use. In the limit of very high SNR, we can approach this data rate with a small probability of errors. However, to approach the performance predicted by the Shannon limit that we discussed in Chapter 3, we have to change our approach and include error correcting codes (discussed in Chapter 8). To be clear, in this scenario the data rate is four times the bandwidth, which reinforces the notion that bandwidth is not data rate.[1]

The BPSK modulation simply switches the modulating voltage's sign at complex baseband. Because the modulation only uses the real part of the signal, only the cosine component of the up-conversion chain seen in Figure 5.5 is employed. In general, because the propagation channel will rotate the constellation (viewed at complex baseband), the receiver will observe both real and imaginary components. The effect of this rotation can be fixed easily at the receiver. In Figure 6.3, we display the carrier signal, the BPSK modulation, and the modulated signal.

We can extend the discussion to QPSK by recognizing that QPSK can be constructed from scaled independent BPSK modulations along the real and imaginary components of the complex baseband plane. We can further extend this discussion to higher-order constellations by allowing a range of amplitudes along the real and imaginary components.

Higher-order modulations employ more complicated distributions of points on the constellation. The QAM modulations employ sets of voltages along both the real and imaginary directions. The PSK modulations place a number of points on the unit circle.

6.1.3   Modulation-Specific Capacity

From Section 3.2.1, expressed in Equation (3.20), we know that the channel capacity is given by

$$c = \max_{p(X)} I(X; Y) \tag{6.10}$$

for output $Y = X + W$ with input signal X in additive noise W. If, instead of a Gaussian source distribution, we consider a different discrete modulation, the mutual information is modified slightly from that we have discussed previously.

---

[1] If someone implies that data rate is bandwidth, smile calmly and back away slowly. They could be unhinged, or even a computer scientist.

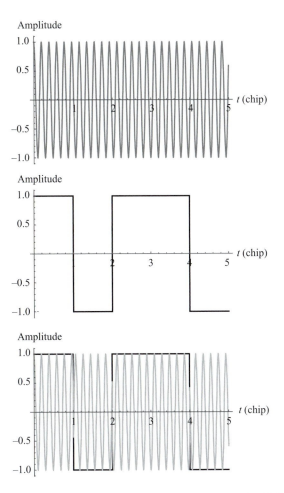

Figure 6.3 These graphs depict a simple example of a BPSK modulation. Respectively, the graphs represent a chip rate-normalized frequency of 5.17, a sequence of BPSK symbols $\{1, -1, 1, 1, -1\}$, and the resulting up-converted signal with the BPSK symbols shown for reference.

Under the assumption of additive complex Gaussian noise (introduced in Section 2.2.4) with variance σ^2, the probability and probability density functions for a set of discrete modulation symbols (which can be represented by a constellation) are given by

$$Q(k) = q_k \tag{6.11}$$

$$p(y|k) = \frac{1}{\pi \sigma^2} e^{-\|y - m_k\|^2 / \sigma^2} \tag{6.12}$$

$$p(y, k) = Q(k)\, p(y|m_k) = q_k\, \frac{1}{\pi \sigma^2}\, e^{-\|y - m_k\|^2 / \sigma^2} \tag{6.13}$$

$$p(y) = \sum_k p(y, k), \tag{6.14}$$

where $Q(k)$ is the probability of the kth symbol being employed with probability value q_k. The actual complex value for the kth symbol is indicated by m_k. We use $p(y|k)$ to indicate the probability density of observing an output of y given that the kth symbol was transmitted. We use $p(y)$ to indicate the probability density of observed signal y. Given these probability definitions, the mutual information is given by

$$I = \int d^2y \sum_k p(y,k) \log_2\left[\frac{p(y,k)}{Q(k)\,p(y)}\right]$$

$$= \sum_k \int d^2y\, Q(k)\, p(y|k) \log_2\left[\frac{Q(k)\,p(y|k)}{Q(k)\,p(y)}\right]$$

$$= \sum_k Q(k) \int d^2y\, p(y|k) \log_2\left[\frac{p(y|k)}{p(y)}\right], \tag{6.15}$$

where d^2y is a shorthand for dy_{Re} and dy_{Im} for the real and imaginary components of y. By assuming that we have circularly symmetric Gaussian noise, the mutual information is given by

$$I = \sum_k q_k \int d^2y\, \frac{1}{\pi\sigma^2} e^{-\|y-m_k\|^2/\sigma^2} \log_2\left[\frac{\frac{1}{\pi\sigma^2} e^{-\|y-m_k\|^2/\sigma^2}}{\sum_j q_j \frac{1}{\pi\sigma^2} e^{-\|y-m_j\|^2/\sigma^2}}\right]$$

$$= \sum_k q_k \int d^2y\, \frac{1}{\pi\sigma^2} e^{-\|y-m_k\|^2/\sigma^2} \log_2\left[\frac{e^{-\|y-m_k\|^2/\sigma^2}}{\sum_j q_j e^{-\|y-m_j\|^2/\sigma^2}}\right]$$

$$= \sum_k q_k \int d^2x\, \frac{1}{\pi\sigma^2} e^{-\|x\|^2/\sigma^2} \log_2\left[\frac{e^{-\|x\|^2/\sigma^2}}{\sum_j q_j e^{-\|x-m_j+m_k\|^2/\sigma^2}}\right] \quad ; x = y - m_k$$

$$= -\sum_m q_m E_x\left\{\log_2\left[\sum_j q_j e^{-\|x-a_j+a_k\|^2/\sigma^2+\|x\|^2/\sigma^2}\right]\right\}. \tag{6.16}$$

A common assumption employed in communications systems is that all points on the constellation are equally likely. Under this assumption, so that $q_m = q_n = 1/N$, the mutual information reduces to

$$I = -\sum_k \frac{1}{N} E_x\left\{\log_2\left[\sum_j \frac{1}{N} e^{-\|x-m_j+m_k\|^2/\sigma^2+\|x\|^2/\sigma^2}\right]\right\}$$

$$= \log_2[N] - \frac{1}{N}\sum_k E_x\left\{\log_2\left[\sum_j e^{-\|x-m_j+m_k\|^2/\sigma^2+\|x\|^2/\sigma^2}\right]\right\}. \tag{6.17}$$

While these integrals are not amenable to analytic evaluation, we can use numerical integration. In Figure 6.4, we compare capacity and the capacity evaluated numerically under the constraints for BPSK, QPSK, 8-PSK, and 16-QAM. Here, the SNR is for the total baseband received signal power divided by the total complex baseband noise power σ^2.

For modulations that are constrained to the real axis, such as BPSK, the imaginary noise component does not cause any potential confusion as it can be removed theoretically without affecting the signal. As a result, there are two valid possible definitions for SNR. For the performance that we depict in Figure 6.4, all the complex noise power is used in evaluating the SNR. Conversely, if we consider only the noise along the real axis, the performance requires a higher real-only SNR by 3 dB to achieve the same performance. To understand this 3 dB shift, consider the total noise power for complex symbols. This power is given by $\sigma^2 = \sigma_{re}^2 + \sigma_{im}^2$, where the subscripts re and im respectively indicate the contributions along the real and imaginary axes of the baseband complex plane. If we know that the symbol is constrained to lie along the real axis, then we can ignore the noise along the imaginary axis, so the noise power that we need to consider is reduced by a factor of two or, equivalently, 3 dB.

In the limit of low SNR, all modulations converge to the true capacity. In the limit of high SNR, the modulation-specific capacities are limited by the number of bits that can be specified by each symbol. An equivalent calculation can be performed for real channels with amplitude-only modulation.

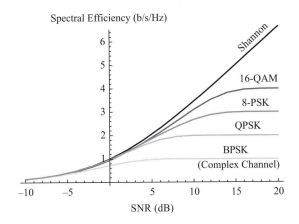

Figure 6.4 Capacity comparison of Gaussian, 16-QAM, 8-PSK, QPSK, and BPSK.

6.1.4 Differential Modulation

While the constellations discussed in Section 6.1.2 employ a fixed set of phases and amplitudes to encode information, one can also use the difference between these constellation points. This differential approach has the advantage of reducing the reliance on a reference phase to demodulate because only the relative complex amplitude is used. Furthermore, by reducing the need for the carrier reference phase, the modulation becomes less sensitive to local oscillator or Doppler-frequency offsets. The reduced sensitivity to frequency offsets comes at the cost of reduced modulation performance because the noise of both the current and previous sample affects the effective SNR for demodulation.

6.1.5 Frequency-Shift Keying

Frequency-shift keying (FSK) is a slight modification to the complex amplitude approach that we presented in Section 6.1.2. In this approach, symbols are encoded by setting the instantaneous frequency (which is the derivative of the phase with respect to time) during the symbol. By considering the complex baseband signal, one can imagine that multiple bits can be represented by selecting various complex tones. An important advantage of this approach is that the complex amplitude at baseband is constant modulus. By keeping the signal a constant magnitude without zero-crossings at complex baseband, the linearity requirements of the hardware (for example, amplifiers) are significantly reduced.

A useful version of this modulation is to enforce that the phase is continuous across symbols; only the phase ramp, which produces the instantaneous frequency shift, is changed. If this is done, the modulation is termed a continuous-phase frequency-shift keying (CPFSK). In practice, it is common to use this approach when encoding a single bit, so that, depending upon the bit, the sign of the derivative of the phase ramp flips. Of particular importance is minimum-shift keying (MSK). This is an FSK modulation defined such that the phase rotates $\pm\pi/2$, or $\pm 1/4$ of the way around the unit circle during each symbol. The baseband instantaneous frequency of each symbol is given by

$$f_\pm = \pm\frac{\alpha}{T}$$

$$= \pm\frac{1}{4T}; \quad \alpha = 1/4, \tag{6.18}$$

where α is the fraction of the unit circle traveled during the symbol.

The MSK modulation gets its name because it is the minimum frequency difference that enables orthogonality between symbols under the assumption that the receiver integrates over the two frequencies at the passband. We can see this by considering the functional inner product between the complex tones of two frequencies over the duration of a symbol. The two complex baseband tones are given by

$$s_{pb,\pm}(t) = \cos(2\pi\,(f_c + f_\pm)\,t + \phi_{m-1}); \quad 0 \le t \le T, \tag{6.19}$$

where the origin of our time axis is set so that the start of the current symbol is at $t = 0$, ϕ_{m-1} is the ending phase of the previous symbol, and f_c is the carrier frequency. If we evaluate the functional inner product between the two passband frequency options $f_c + f_\pm$, integrated over a symbol duration, we see that the symbol-duration-normalized inner product approaches zero:

$$\mathcal{I} = \frac{1}{T}\langle s_{pb,-}(t), s_{pb,+}(t)\rangle$$

$$= \frac{1}{T}\int_0^T dt\, \cos(2\pi\,[f_c + \alpha/T]\,t + \phi_{m-1})\, \cos([2\pi\,(f_c - \alpha/T)\,t + \phi_{m-1}])$$

$$= \frac{1}{T}\frac{2\alpha\,\sin(2\pi f_c T)\,\cos(2[\pi f_c T + \phi]) + f_c T\,\sin(4\pi\alpha)}{8\pi\alpha f_c}$$

$$
\begin{aligned}
&= \frac{2(1/4)\sin(2\pi f_c T)\cos(2\pi f_c T + 2\phi) + f_c T \sin(4\pi [1/4])}{8\pi (1/4) f_c T} \\
&= \frac{\sin(2\pi f_c T)\cos(2\pi f_c T + 2\phi)}{4\pi f_c T} \approx 0
\end{aligned} \tag{6.20}
$$

because $f_c T$ is typically quite large. Because, in a modern digital system, we would coherently sample the phase at the appropriate time and determine the symbol accordingly, the argument above may seem a little strange. Nonetheless, the MSK modulation is useful in practice by including intermediate points at a sample rate higher than the chip rate.

6.1.6 Gaussian Minimum-Shift Keying

An interesting extension to the MSK modulation is the Gaussian minimum-shift keying (GMSK) modulation. We often need to shape the spectrum of the modulation. Most commonly we use amplitude pulse-shaping approaches that we discuss in Section 6.2.2. However, it is possible to filter the phase of a PSK or FSK modulation. For example, we can apply a filter whose taps are weighted by a shape given by a Gaussian shape to the phase of an MSK modulation. By filtering the phase rather than the amplitude, we maintain the constant modulus waveform. However, these filters are less effective in suppressing out-of-band spectral sidelobes.

6.2 Spectrum and Pulse Shaping

We often assume that the required bandwidth for a modulation is given by the inverse of the symbol (or chip) rate. In general, this is an optimistic assumption.

6.2.1 Spectral Shape

As we discuss in Section 15.4.13, we can find the power spectral density of a random process $x(t)$ by evaluating the Fourier transform of the autocorrelation $r_x(\tau)$. For this approach to be valid, we assume that our signal is drawn from a wide-sense stationary random process (see Section 14.2.3). From Equations (15.48) and (15.67), we have the power spectral density $P(f)$ given by

$$
P(f) = \mathcal{F}\{r_x(\tau)\}
$$
$$
r_x(\tau) = E\{x(t)\,x^*(t - \tau)\}, \tag{6.21}
$$

where the complex baseband transmitted sequence $x(t)$ for the nth modulation symbol m_n is given by

$$
x(t) = \sum_{n=-\infty}^{\infty} m_n T_s \, \Pi(t; T_s) \tag{6.22}
$$

and the top hat function is defined by

$$\Pi(t; T_s) = \begin{cases} \frac{1}{T_s} & ; \quad \|t\| \leq T_s/2 \\ 0 & ; \quad \text{otherwise.} \end{cases} \tag{6.23}$$

The power is defined to be

$$E[\|x(t)\|^2] = E[\|m_n\|^2] = \sigma^2. \tag{6.24}$$

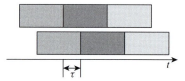

Figure 6.5 Example of three symbols offset for τ that are used by autocorrelation evaluation.

For an example modulation discussed here, it is assumed that the signal is constant over a symbol duration of length T_s, as depicted for a notional sequence of symbols and symbols offset by delay τ in Figure 6.5. The symbol then changes abruptly to the signal's next value. To evaluate Equation (6.21), we observe that each symbol is independent, and we assume that the symbols have zero mean. Consequently, the only offsets that are nonzero are when the delay τ is sufficiently small so that each symbol overlaps with itself. The overlap is maximized when the delay is zero. The autocorrelation falls linearly as the delay increases (or decreases) until the end of the symbol is found. The autocorrelation is then given by

$$r_x(\tau) = E\{x(t) x^*(t - \tau)\}$$

$$= \sigma^2 \left(1 - \frac{\|\tau\|}{T_s}\right) \quad \forall \|\tau\| \leq T_s, \tag{6.25}$$

where σ^2 is the variance or power of the signal. We have noted that the triangle shape of $r_x(\tau)$ can be constructed from the convolution of two top hat functions:

$$\Pi(t; T_s) = \begin{cases} \frac{1}{T_s} & ; \quad \|\tau\| \leq T_s \\ 0 & ; \quad \text{otherwise} \end{cases}$$

$$\Pi(\tau; T_s) * \Pi(\tau; T_s) = \int dt \, \Pi(t; T_s) \, \Pi(\tau - t; T_s)$$

$$= T_s \frac{1}{T_s^2} \left(1 - \frac{\|\tau\|}{T_s}\right) ; \quad \|\tau\| \leq T_s$$

$$\left(1 - \frac{\|\tau\|}{T_s}\right) = T_s \, \Pi(\tau; T_s) * \Pi(\tau; T_s). \tag{6.26}$$

Because the Fourier transform of the convolution of functions is the product of the Fourier transforms of the functions, we can quickly evaluate Equation (6.21) using the observation that the Fourier transform of a top hat function is a sinc function. Consequently, Equation (6.21) evaluates to

$$P(f) = \mathcal{F}\{r_x(\tau)\}$$
$$= \sigma^2 \, T_s \, \text{sinc}^2(f \, T_s), \tag{6.27}$$

where $\text{sinc}(f) = \sin(\pi f)/(\pi f)$. The resulting power spectral density is depicted in Figure 6.6. We observe that the power is not well confined in the spectrum. There is significant power outside the bandwidth that we would expect based upon the chip rate $\pm 1/(2T_s)$ or $\pm 1/2$ for normalized frequencies. This poor spectral confinement motivates other techniques, such as pulse shaping.

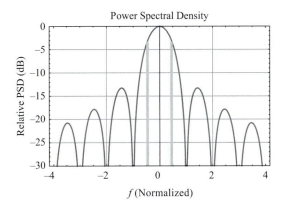

Figure 6.6 Power spectral density of a QPSK signal with top hat pulse shaping compared to the expected spectral confinement based upon a total bandwidth defined by $1/T_s$.

6.2.2 Pulse-Shaping Filter

We can consider a sequence of source symbols $\tilde{s}(t)$ as a sequence of amplitudes m_k selected from the modulation's constellation multiplied by a delta function at the appropriate point in time, which is given by

$$\tilde{s}(t) = \sum_{k=-\infty}^{\infty} m_k \, \delta(t - k \, T_s) \tag{6.28}$$

for some sample period T_s. We convolve the pulse shape $g(t)$ with this function to provide the actual transmit sequence, which is given by

$$s(t) = [g * \tilde{s}](t)$$
$$= \int d\tau \, g(\tau) \sum_{k=-\infty}^{\infty} m_k \, \delta(t - \tau - k \, T_s)$$
$$= \sum_{k=-\infty}^{\infty} m_k \int d\tau \, g(\tau) \, \delta([t - k \, T_s] - \tau)$$
$$= \sum_{k=-\infty}^{\infty} m_k \, g(t - k \, T_s). \tag{6.29}$$

For the discussion in Section 6.2.1, it was implicitly assumed that the pulse-shaping filter was a top hat with width of the symbol or chip duration T_s. We can use other pulse-shaping filters to help constrain the width of the spectrum.

The resulting transmitted power spectral density is a function of the spectral shape of the pulse-shaping filter. We use Equations (15.67) and (15.48) to evaluate the power spectral density of the filtered signal. The autocorrelation of $s(t)$ is given by

$$r_s(\tau) = E\{s(t)\, s^*(t - \tau)\}$$

$$= E\left\{\left(\sum_{k=-\infty}^{\infty} m_k\, g(t - k\, T_s)\right)\left(\sum_{j=-\infty}^{\infty} m_j\, g(t - \tau - j\, T_s)\right)^*\right\}. \qquad (6.30)$$

By assuming m_k is zero mean, unit variance, and independent such that $E[m_k m_j^*] = \delta_{k,j}$, we have only terms involving the same index,

$$r_s(\tau) = E\left\{\sum_{k=-\infty}^{\infty} g(t - k\, T_s)\, g^*(t - \tau - k\, T_s)\right\}$$

$$= \lim_{n\to\infty} \frac{1}{(2n+1)\, T_s} \int_{-nT_s}^{nT_s} dt \sum_{k=-n}^{n} g(t - k\, T_s)\, g^*(t - \tau - k\, T_s)$$

$$= \lim_{n\to\infty} \frac{1}{(2n+1)\, T_s} (2n+1) \int_{-nT_s}^{nT_s} dt\, g(t)\, g^*(t - \tau)$$

$$= \frac{1}{T_s} \int_{-\infty}^{\infty} dt\, g(t)\, g^*(t - \tau), \qquad (6.31)$$

where we have observed that we can move the origin of integration when integrating over the entire axis. The power spectral density is then given by

$$\mathcal{F}\{r_s(\tau)\} = \int_{-\infty}^{\infty} d\tau\, e^{-i\,2\pi\,\tau f}\, r_s(\tau)$$

$$= \int_{-\infty}^{\infty} d\tau\, e^{-i\,2\pi\,\tau f}\, \frac{1}{T_s} \int_{-\infty}^{\infty} dt\, g(t)\, g^*(t - \tau)$$

$$= \frac{1}{T_s} \int_{-\infty}^{\infty} dt\, g(t) \int_{-\infty}^{\infty} d\tau'\, e^{i\,2\pi\,(\tau'-t)f}\, g^*(\tau')$$

$$= \frac{1}{T_s} \int_{-\infty}^{\infty} dt\, e^{-i\,2\pi\,tf}\, g(t) \int_{-\infty}^{\infty} d\tau'\, e^{i\,2\pi\,\tau' f}\, g^*(\tau')$$

$$= \frac{1}{T_s} \int_{-\infty}^{\infty} dt\, e^{-i\,2\pi\,tf}\, g(t) \left[\int_{-\infty}^{\infty} d\tau'\, e^{-i\,2\pi\,\tau' f}\, g(\tau')\right]^*$$

$$= \frac{1}{T_s} G(f)\, G^*(f) = \frac{1}{T_s} \|G(f)\|^2, \qquad (6.32)$$

where $G(f) = \mathcal{F}\{g(t)\}$. Thus, the power spectral density after pulse shaping is proportional to the magnitude squared of the Fourier transform of the pulse-shaping filter.

For perfect spectral shaping with a top hat shaped spectrum, we can use the fact that convolution corresponds to multiplication in the frequency domain. If we multiply the raw modulation spectrum by a top hat, then we can achieve this spectral shaping by convolving the signal with the inverse Fourier transform of the top hat $G(f)$ with the bandwidth given by the inverse of the symbol period $1/T_s$, which is given by a sinc function. Thus, we use a pulse-shaping function $g(t)$ that is given by the inverse Fourier transform of the top hat function, which is given by

$$g(t) \propto \mathcal{F}^{-1}\{G(f\,T_s)\} = \frac{\sin(\pi t/T_s)}{\pi\, t/T_s}, \tag{6.33}$$

as depicted in Figure 6.7. The implementation of the filter is performed by using a sampled version of $g(t)$, where $t = k\,T_s$. It is worth noting that the sample period might be smaller than the period of T_s. Many real systems operate in an oversampled regime. In this regime, the digital samples are typically by some easy multiple (such as 2 or 4) of the symbol rate.

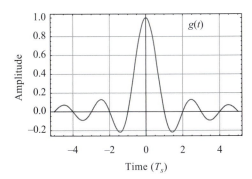
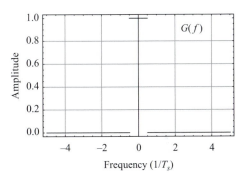

Figure 6.7 Correspondence between spectral top hat and sinc pulse shaping.

Conveniently, the sinc function has a particularly useful characteristic of orthogonality if sampled at the correct point in time, as depicted in Figure 6.8. The zeros of all time-delayed sinc functions are aligned so that they are zero at the peak of each symbol's zero delay offset. In dispersive channels, which are discussed in Chapter 7, this orthogonality will be broken by delay spread induced by the channel, although the bandwidth-limiting effect is maintained.

A significant difficulty with this pulse-shaping approach is that it technically requires the knowledge of a symbol before the formation of the universe. This non-causal requirement can be easily fixed by truncating the sinc to be limited in temporal extent; however, this truncation breaks the perfect spectral confinement.

In practice, there are a number of approaches that trade pulse-shaping filter length for spectral confinement. Commonly used approaches include raised-cosine pulse shaping and root-raised-cosine pulse shaping [10]. In the frequency domain (assuming some Nyquist symbol period T_s), the raised-cosine filter is given by

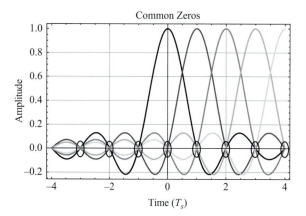

Figure 6.8 Number of sinc shaped symbol pulses that are offset by T_s in time.

$$G_{rc}(f) = \begin{cases} 1 & ; \quad \|f\| \le \frac{1-\beta}{2T_s} \\ \frac{1}{2}\left[1 + \cos\left(\frac{\pi T}{\beta}\left[\|f\| - \frac{1-\beta}{2T_s}\right]\right)\right] & ; \quad \frac{1-\beta}{2T_s} < \|f\| \le \frac{1+\beta}{2T_s}, \\ 0 & ; \quad \text{otherwise} \end{cases} \quad (6.34)$$

where the pulse-shaping filter is parameterized by rolloff factor β. The raised-cosine pulse-shaping filter in time is then given by

$$g_{rc}(t) = \frac{\sin(\pi t/T_s)}{\pi t/T_s} \frac{\cos(\pi \beta t/T_s)}{1 - 4\beta^2 t^2/T_s^2}. \quad (6.35)$$

The root-raised-cosine filter $G_{rrc}(f)$ in the frequency domain is given by

$$G_{rrc}(f) = \sqrt{G_{rc}(f)}. \quad (6.36)$$

This root-raised-cosine filter has the advantage in that, if the filter is applied by the transmitter and then by the receiver, the overall effect is simply the raised-cosine filter that preserves the orthogonality of successive symbols. The root-raised-cosine pulse-shaping filter [10] in time is then given by

$$g_{rrc}(t) = \lim_{\tau \to t} \frac{1}{T_s} \frac{\sin\left[\pi \frac{\tau}{T_s}(1 - \beta)\right] + 4\beta \frac{\tau}{T_s} \cos\left[\pi \frac{\tau}{T_s}(1 - \beta)\right]}{\pi \frac{\tau}{T_s}\left[1 - \left(4\beta \frac{\tau}{T_s}\right)^2\right]}. \quad (6.37)$$

Interestingly, the pulse-shaping filter does not have zeros at $t = kT_s$ for $k \in \mathbb{Z}$. However, when the receive pulse-shaping filter is applied so that the effect of both filters is $G_{rc}(f) = G_{rrc}(f)\,G_{rrc}(f)$, the zeros return.

The raised-cosine and root-raised-cosine filters have limited spectral support, so we would expect the filter to be of infinite length. We quickly recognize that this is not practical, and we will need to truncate the filter to use it. This truncation degrades the ideal spectral characteristics. In practice, we have to find the right number of taps to achieve the spectral performance that the system needs.

In addition, it is worth stressing that channel dispersion (that is, channels that have echos caused by channel delay spread) breaks the symbol orthogonality, reducing the benefits of root-raised-cosine filters. For mild dispersion, a common approach is to first use some pulse-shaping filter at the transmitter to constrain the spectrum and then to use an adaptive equalizer in the receiver to improve performance. For channels with significant dispersion, other approaches are often employed, such as orthogonal frequency division multiplexing (OFDM). These approaches are discussed in Chapter 7.

6.3   Demodulation

6.3.1   Flat-Fading Channel Compensation

The effect of a flat-fading channel is to attenuate and rotate the signal and then add Gaussian noise. At complex baseband, we represent that amplitude attenuation magnitude and phase rotation with complex parameter a and associated estimate $\hat{a}$. The magnitude of a is much smaller than 1 because most of the transmitted power does not arrive at the receiver. The complex phase is motivated by the observation that delays at passband for a narrowband signal translate into a change in phase at complex baseband.

The standard complex baseband model for a given symbol in flat-fading is given by

$$z = a\,s + n,\tag{6.38}$$

where z is the observed signal, s is the transmitted symbol, and n is the Gaussian noise. To prepare for decoding, we compensate for the effect of the channel. There are, in general, two approaches. You can either modify the observed signal to compensate for the channel or modify the model to match the observed signal. Here, we will modify the observed signal by rotating and scaling the observed signal, as indicated in Figure 6.9. If we have an observation, we can construct an estimate of the transmitted signal by multiplying the observed signal by the inverse of the estimated channel. The estimate of the transmitted symbol is given by

$$\hat{s} = \hat{a}^{-1}\,z = \hat{a}^{-1}\,a\,s + \hat{a}^{-1}\,n.\tag{6.39}$$

If the channel estimate is perfect $\hat{a} = a$, then the symbol estimate is given by

$$\hat{s} = s + a^{-1}\,n.\tag{6.40}$$

It is sometimes useful to represent the inverse with the form

$$a^{-1} = \frac{a^*}{\|a\|^2}.\tag{6.41}$$

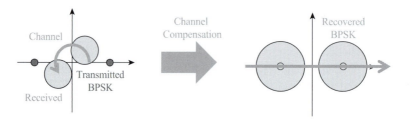

Figure 6.9 Channel compensation to rotate and scale the BPSK signal after attenuation and rotation caused by the flat-fading channel.

6.3.2   BPSK Demodulation

Under the assumption that the channel has been corrected by the techniques discussed in Section 6.3.1, the demodulation of a BPSK symbol can be determined if we consider the distribution of observed signals in Gaussian noise, as observed in Figure 6.10 for a 0 dB SNR real Gaussian channel. The two probability density functions are associated with BPSK symbols $s = \pm 1$, respectively. If one had to make a hard decision at this point (which is wildly suboptimal), then it is clear that the best you can do is to set a threshold at $z = 0$ so that $z > 0$ implies $\hat{s} = 1$ and $\hat{s} = -1$ otherwise.

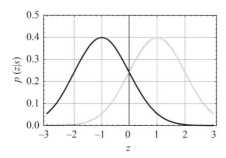

Figure 6.10 Probability density function for observation variable z for BPSK ($s = \pm 1$) in the presence of real Gaussian noise of unit variance.

Employing this threshold is called *making a hard decision* or *hard demodulation*. As we discussed in the development of the theoretical capacity presented in Chapter 3, the capacity is found by decoding a large vector of symbols at once, so we know that making hard decisions symbol by symbol must be suboptimal. By making decisions one symbol at a time, it is impossible to get to error-free performance with finite SNR. For a real channel with BPSK modulation, we can calculate the probability of error. If the BPSK symbols are ± 1 with unit noise variance and real amplitude a, the probability density function for an observed signal,

$$z(t) = a\, s(t) + n(t), \tag{6.42}$$

is given by

$$p(z|s = -1, a) = \frac{1}{\sqrt{2\pi}} e^{-(z - a[-1])^2/2} , \qquad (6.43)$$

under the assumption that the noise variance is unity. With this parameterization, the SNR is given by $\|a\|^2$. Given that symbols ± 1 are equally likely, the logical threshold is 0. An error for a decision made by itself occurs if the observed signal fluctuates above 0. This error P_{error} is given by

$$
\begin{aligned}
P_{\text{error}} &= \int_0^\infty dz\, p(z|s = -1, a) \\
&= \int_0^\infty dz\, \frac{1}{\sqrt{2\pi}} e^{-(z - a[-1])^2/2} \\
&= Q(a) ,
\end{aligned}
\qquad (6.44)
$$

where the Gaussian Q-function is defined by

$$Q(a) = \int_a^\infty dz\, \frac{1}{\sqrt{2\pi}} e^{-z^2/2} . \qquad (6.45)$$

In Figure 6.11, we demonstrate this difference in performance between the Shannon limit for a 1 b/s/Hz spectral efficiency in a real channel,

$$1 = \frac{1}{2} \log_2(1 + \text{SNR})$$

$$\text{SNR} = 2^2 - 1 \approx 4.8\,\text{dB} , \qquad (6.46)$$

and a symbol-level decision for BPSK.

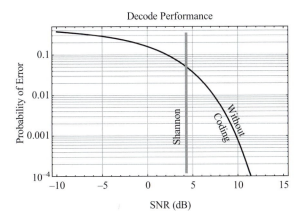

Figure 6.11 Comparing simple hard decoder versus channel capacity, bit error rate as a function of SNR for the BPSK signal in real Gaussian noise.

6.3.3   QPSK Demodulation

We now consider the bit error rate (BER) and the symbol error rate (SER) of QPSK. To maintain unit variance transmit sequence, the points on the constellation are closer together, so errors are more likely. However, because the number of points in the constellation is greater, it is possible to send data at a higher rate. The possible symbol locations in the complex plane are represented by $(\pm 1 + \pm i)/\sqrt{2}$. By considering the real and imaginary positions as modulating independent bits, we observe that the bit error rate can be calculated from the BPSK calculation. The amplitude along the real or imaginary direction is reduced by $\sqrt{2}$. Similarly, the variance of noise along each direction is $1/2$ of the total noise, so the standard deviation is reduced by $\sqrt{2}$. Consequently, the probability of error is given by

$$P_{r,\text{error}} = P_{i,\text{error}} = Q\left(\frac{a/\sqrt{2}}{1/\sqrt{2}}\right) = Q(a). \tag{6.47}$$

To evaluate the SER, both components in the complex plane must be identified correctly. Because the distances to all other points in the complex plane are independent of whichever symbol is selected, we can calculate the symbol error based on a single reference choice. For more complicated modulations, this may not be true. In Figure 6.12, we depict the region for correct symbol identification.

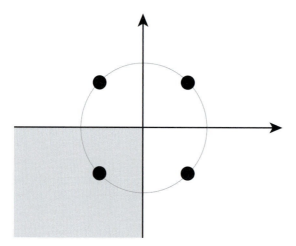

Figure 6.12 Region of correct decision for $-(1 + i)/\sqrt{2}$ symbol.

If the probability of an error or correct decision in the real or imaginary direction is indicated by $P_{r,\text{error}}$, $P_{r,\text{cor}}$, $P_{i,\text{error}}$, and $P_{i,\text{cor}}$, respectively, then we can determine the probability of symbol error by rewriting 1 in the following way:

$$1 = P_{r,\text{error}} + P_{r,\text{cor}} = P_{i,\text{error}} + P_{i,\text{cor}}$$

$$= (P_{r,\text{error}} + P_{r,\text{cor}})(P_{i,\text{error}} + P_{i,\text{cor}})$$

$$P_{\text{sym error}} = 1 - P_{r,\text{cor}} P_{i,\text{cor}}$$
$$= 1 - [1 - P_{r,\text{error}} P_{i,\text{cor}} - P_{r,\text{cor}} P_{i,\text{error}} - P_{r,\text{error}} P_{i,\text{error}}]$$
$$= P_{r,\text{error}}(1 - P_{i,\text{error}}) + (1 - P_{r,\text{error}})P_{i,\text{error}} + P_{r,\text{error}} P_{i,\text{error}}$$
$$= 2 P_{r,\text{error}} - P_{r,\text{error}}^2, \tag{6.48}$$

where we have observed that $P_{r,\text{error}} = P_{i,\text{error}}$ for this modulation and noise assumption. The overall SER of QPSK $P_{\text{sym error}}$ is given by

$$P_{\text{sym error}} = 2 Q(a) - Q^2(a), \tag{6.49}$$

as is depicted in Figure 6.13. For reference, we include the required SNR to achieve ideal performance of 2 b/s/Hz that QPSK ideally provides:

$$2 = \log_2(1 + \text{SNR})$$
$$\text{SNR} = 3 \approx 4.8 \text{ dB} \tag{6.50}$$

It is clear that simply making decisions one symbol at a time is suboptimal. At low SNR, the probability of symbol error approaches 3/4 because either bit error causes a symbol error. At high SNR, the error is proportional to $P_{\text{sym error}} \propto e^{-\text{SNR}/2}$.

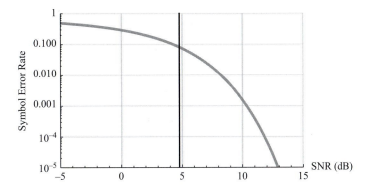

Figure 6.13 Comparison of SNR required to achieve 2 b/s/Hz from information theoretic capacity and QPSK symbol error rate.

As one can imagine, more complicated modulation constellations require more sophisticated calculations for BER and SER. However, with modern error correcting codes, these raw BERs can be misleading. Consequently, we do not stress them here.

6.4   Demodulation Likelihoods

In modern communications, it is uncommon to demodulate one symbol at a time. It is far more common to evaluate a likelihood of one symbol or another and to provide these likelihoods to the decoder of an error correcting code. This approach enables much better performance. Often, the logarithm of the likelihood (log likelihoods) is used because it requires a smaller numerical range. In Gaussian noise, this likelihood can be evaluated by measuring the Cartesian distance from the observed signal and potential symbols of the constellation. The log likelihood for the mth symbol for an observation of z is given by

$$ll(m) = \log_e[p(z|s_m)] = k - \frac{\|z - \hat{a}\, s_m\|^2}{\sigma^2}, \qquad (6.51)$$

where k is the logarithm of the constant terms, $\hat{a}$ is the estimated channel, and σ^2 is the complex noise as variance. In special cases for which the constellation enables symbols with binary choices, the logarithm of the likelihood ratios is employed. For example, when using a BPSK symbol, we often consider the log-likelihood ratio $ll(1) - ll(0)$, which is simply the difference of the two possible log likelihoods.

6.5   Flat-Fading Channel Coefficient Estimation

We assume an n_s sequence of transmitted training (or pilot) symbols $\underline{s} \in \mathbb{C}^{1 \times n_s}$ that are known at both the transmitter and receiver. We denote the observed signal at the receiver at complex baseband $\underline{z} \in \mathbb{C}^{1 \times n_s}$ under the model of attenuation-gain product $\|a\|^2$ and circular symmetric Gaussian noise. The resulting probability density function for the observed signal $\underline{z}$ is given by

$$p(\underline{z}) = \frac{1}{\pi^{n_s}\, \sigma^{2n_s}}\, e^{-(\underline{z}-a\,\underline{s})(\underline{z}-a\,\underline{s})^H/\sigma^2}. \qquad (6.52)$$

We can find the maximum likelihood estimate of parameter a by finding the value of a that maximizes the probability density function for some observed signal $\underline{z}$ and given training sequence $\underline{s}$. By employing the optimization approach that is explained in Section 13.8, we express the complex parameters a as a function of real parameter α and find the maximum likelihood estimator for parameter a:

$$a = a(\alpha)$$
$$0 = \frac{\partial}{\partial \alpha} \log_e[p(\underline{z}; a[\alpha], a^*[\alpha])]$$
$$= (\underline{z} - a\,\underline{s})(\underline{s})^H \left(\frac{\partial}{\partial \alpha}\right) \frac{1}{\sigma^2} + h.c.$$
$$0 = (\underline{z} - a\,\underline{s})(\underline{s})^H$$
$$\hat{a} = \underline{z}\,\underline{s}^H\,(\underline{s}\,\underline{s}^H)^{-1}, \qquad (6.53)$$

where we use $h.c.$ to indicate the Hermitian conjugate of the first term. We use the notation $\hat{\cdot}$ to identify the estimate $\hat{a}$ as a separate object because, in general, $\hat{a} \approx a$.

Problems

6.1 Identify the number of bits required to identify a point on the constellation for
(a) BPSK
(b) 64-QAM
(c) 8-PSK

6.2 Consider complex QPSK, 8-PSK, and 16-QAM baseband modulations. By using simulation, draw and plot a sequence of 100 random symbols for unit-variance noise power and SNR 10 dB.
(a) Plot the resulting constellation for each modulation.
(b) Apply a normalized frequency shift of 1/800 and plot the constellations in the absence of noise.

6.3 Consider complex QPSK and 8-PSK baseband modulations. Calculate the values of SNR to correctly identify constellation points with an average likelihood of 0.9.

6.4 Consider symbols with symmetric triangle pulse shaping of duration T and symbol spacing T. Evaluate the shape of the power spectral density of the pulse-shaped symbols.

6.5 In simulation, construct a sequence of 1000 random QPSK symbols each repeated 10 times successively, which corresponds to a top hat pulse that is 10-times oversampled. Plot the periodigram estimate of the power spectral density.

6.6 In simulation, construct a sequence of 1000 random QPSK symbols. Employ a five-sample, truncated sinc pulse-shaping filter and plot the periodigram estimate of the power spectral density for a 10-times oversampled signal. Compare to Problem 6.5. Hint: convert to an oversampled signal without pulse shaping first and then introduce pulse shaping.

6.7 Consider a QPSK modulation. Evaluate and plot the likelihood of a (+,+) as a function of observed position in the complex plane under the assumption of SNR of 0 dB.

6.8 By using numerical integration, evaluate the modulation-specific capacity for BPSK, QPSK, and 8-PSK as a function of SNR in the complex plane. Plot over the range of -10 to 20 dB.

6.9 Evaluate and plot the power spectral density for a QPSK with:
(a) a top hat pulse-shaping filter; and
(b) a sinc pulse-shaping filter.

6.10 Plot the power spectral density for root-raised-cosine pulse shaping for values of $\beta = 0.1, 0.5, 1$.

6.11 For a QPSK constellation, compare the probability of symbol error and evaluate for a single symbol as the function of SNR versus that given by the Shannon capacity.

6.12 Using a simulation and given a flat-fading channel with an instantaneous SNR of 0 dB, evaluate an estimate of the channel coefficient given eight pseudorandom QPSK training sequences. Evaluate the variance of the channel estimate.

6.13 Design a communications link that connects from the moon to a low-Earth-orbit satellite. The desired data rate is 50 Mb/s. The carrier frequency is 60 GHz. Assume temperature is maintained at 200 K. The maximum gain of the satellite antenna is

23 dB. The receiver noise figure is 3 dB. Assume that on average for every data bit, we require an additional bit for error correction. It's always nice to have a bit of margin. You are free to choose all other parameters. Specify all required system and modulation parameters. Show that the required data rate is achieved. Calculate approximate effective areas of the antennas. Explain why these are reasonable system design choices.

7   Dispersive Channels

In this chapter, we discuss the effects of channel dispersion, or equivalently, the effects of resolvable multipath, and techniques for enabling communications in these environments. We introduce the model for delay spread that is the source of dispersion. We relate the time-domain and frequency-domain representations of the propagation channel. We introduce the approach of adaptive equalization, including zero-forcing and Weiner filtering. We provide an example of finite-sample Weiner filtering. We also introduce the orthogonal frequency-division multiplexing (OFDM) approach to compensate for dispersive channels. We describe OFDM's processing chain. We determine the waveform characteristics and spacing of the subcarriers used to construct OFDM. Finally, we discuss models for dispersive channels.

When a signal propagates from a transmitter to a receiver through a physical channel, the signal is distorted by the scatterers in the environment. Because the direct path and various bounced paths have different lengths, the signal is distorted by the addition of delayed copies of the signal, as depicted in Figure 7.1. Instead of a single delay introduced by the line-of-sight propagation, there is delay spread. This resulting propagation effect on the channel is sometimes termed *dispersive* or *frequency-selective*. These delays introduce inter-symbol interference (ISI) – that is, echoes of previous symbols interfere with the current symbol. This is a particularly insidious form of distortion because if the signal power is increased the distorting power is increased proportionally.

If the transmitters, receiver, or scatterers are moving with respect to each other, then we say that the channel is dynamic, which means the details of the contributions of delays and scattering are changing. We sometimes identify this channel as having Doppler-frequency spread. It is also worth noting that errors and phase noise in physical oscillators can introduce frequency offsets and frequency spread that mimics Doppler-frequency spread. A channel that has both delay spread and Doppler-frequency spread is said to be doubly dispersive. In this discussion, we ignore the effects of dynamic channels, but we wanted to admit to the existence of these effects.

To compensate for the effects of dispersion, we need to potentially modify multiple blocks in the processing chain. We indicate these blocks in Figure 7.2.

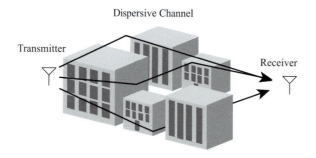

Figure 7.1 Multiple signal paths from transmitter to receiver introduce delay spread and consequently a dispersive channel.

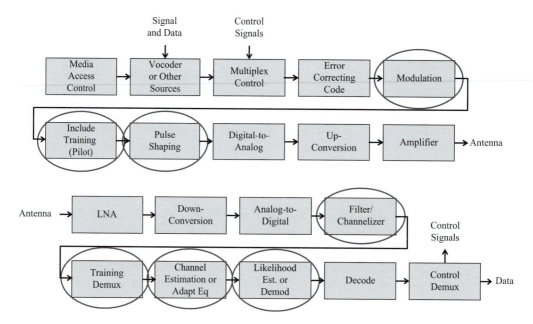

Figure 7.2 Representation of a typical communication system block diagram with blocks affected by dispersion compensation indicated.

7.1   Delay Spread

In general, we can express the effects of scatterers on the channel at complex baseband by adding delayed versions of the signal with arbitrary delays τ_m and amplitudes a_m. The dispersive channel $\tilde{h}(\tau)$ can be represented by

$$\tilde{h}(\tau) = \sum_m a_m \, \delta(t - \tau_m). \tag{7.1}$$

We sometimes denote this $\tilde{h}(\tau)$ as the channel impulse response. The received signal is then given by the convolution of the transmitted signal $s(t)$ and the channel $h(\tau)$ so that the received signal $z(t)$ in some additive Gaussian noise $n(t)$ is given by

$$z(t) = [\tilde{h} * s](t) + n(t)$$

$$= \int d\eta \, s(\eta) \, \tilde{h}(t - \eta) + n(t)$$

$$= \int d\eta \, s(\eta) \sum_m a_m \, \delta(t - \eta - \tau_m) + n(t)$$

$$= \sum_m a_m \int d\eta \, s(\eta) \, \delta([t - \tau_m] - \eta) + n(t)$$

$$= \sum_m a_m \, s(t - \tau_m) + n(t) \,, \tag{7.2}$$

which is unsurprising because one might have started with this form. Given our original form, we can employ a useful observation to rewrite our representation of the channel: If we know that the bandwidth of the transmitted signal is limited, then we can rewrite the channel in a form that is consistent with discretely sampled signals. As an example, if we were to incorporate an ideal low-pass filter represented by a top hat shape denoted by $G(f)$,

$$G(f) = \begin{cases} 1 & ; \, \|f\| \le B/2 \\ 0 & ; \, \text{otherwise} \end{cases}, \tag{7.3}$$

then the temporal representation is $g(t)$, which is proportional to a sinc function and is given by

$$g(t) = B \frac{\sin(\pi \, t \, B)}{\pi \, t \, B} \,. \tag{7.4}$$

If the bandwidth of $S(f) = \mathcal{F}\{s(t)\}$ falls within the shape of the top hat determined by $G(f)$, then

$$S(f) = G(f) \, S(f) \,, \tag{7.5}$$

and therefore

$$s(t) = [g * s](t) \,. \tag{7.6}$$

We can rewrite the original form of the receive signal as

$$z(t) = [\tilde{h} * s](t) + n(t)$$
$$= [\tilde{h} * g * s](t) + n(t)$$
$$= [h * s](t) + n(t) \,, \tag{7.7}$$

where the band-limited representation of the channel $h(\tau)$ is given by

$$h(\tau) = [\tilde{h} * g](\tau)$$

$$= \int d\eta \, g(\eta) \, \tilde{h}(\tau - \eta)$$

$$= \int d\eta \, g(\eta) \sum_m a_m \, \delta(\tau - \eta - \tau_m)$$

$$= \sum_m a_m \int d\eta \, B \, \frac{\sin(\pi \, \eta \, B)}{\pi \, \eta \, B} \, \delta([\tau - \tau_m] - \eta)$$

$$= B \sum_m a_m \frac{\sin(\pi \, [\tau - \tau_m] \, B)}{\pi \, [\tau - \tau_m] \, B} = B \sum_m a_m \, \mathrm{sinc}([\tau - \tau_m] \, B). \tag{7.8}$$

Now both $s(t)$ and $h(\tau)$ are band limited by B, so we can replace the continuous signal with regularly sampled complex signals $s_m = s(m\,T)$ and $h_n = h(n\,T)$ under the constraint that the complex sample rate satisfies the Nyquist criteria $T \leq 1/B$, where T is the time between samples. One benefit of this observation is that we can represent the dispersive channel that has a set of arbitrary delays with a regularly sampled form. On the down side, the sampled form technically requires an infinite number of samples to represent even relatively simple dispersive channels. However, typically it only takes a few samples to provide a sufficiently accurate representation.

By formulating the effects of the channel in this way, we have implicitly assumed that our system is linear time-invariant (LTI). These LTI systems have a number of advantages in terms of analysis and interpretation. As an example, we can meaningfully relate our channel in terms of an impulse response and its corresponding spectral response. However, in all physical systems this is only an approximation, because there is always some nonlinear contribution and no system is completely static. However, we can often employ LTI system assumptions because the approximation is sufficiently accurate to be useful.

7.2   Spectral Effect

As we discuss in Section 15.4.11, the Fourier transform of the convolution of two functions is the product of the Fourier transforms of the two functions. Consequently, the Fourier transform of the received signal is given by

$$\mathcal{F}\{z(t)\} = \mathcal{F}\{[h * x](t) + n(t)\}$$

$$= H(f)\,X(f) + N(f), \tag{7.9}$$

where $H(f)$, $X(f)$, and $N(f)$ are the Fourier transforms of $h(t)$, $x(t)$, and $n(t)$, respectively. A simple channel with a single delay can be represented by

$$h(t) = a \, \delta(t - \tau). \tag{7.10}$$

In the frequency domain, this simple delay channel corresponds to

$$\mathcal{F}\{h(t)\} = \mathcal{F}\{a \, \delta(t - \tau)\} = a \, e^{i \, 2\pi \, \tau f}, \tag{7.11}$$

so the magnitude squared of the channel in the frequency domain in this case is a constant: $\|H(f)\|^2 = \|a\|^2$. Conversely, if there is resolvable delay spread, the channel can vary significantly as a function of frequency, as depicted in Figure 7.3.

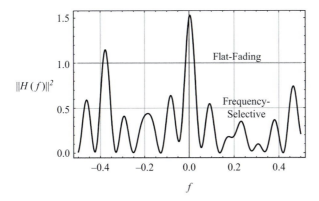

Figure 7.3 Comparison of flat-fading and frequency-selective channels.

7.2.1 Frequency Hopping

One approach to deal with dispersive channels is to employ a moving narrowband signal. Because the signal may be at a frequency that has significant attenuation, we choose to periodically change (or "hop") the center frequency on some predetermined schedule so that both the transmitter and receiver know when these changes will occur. A key benefit to this approach is that the communications link will always get chances with relatively good propagation frequencies. We can use some sort of coding to overcome the occasional bad propagation frequency. Somewhat famously, this approach was invented and patented by actress Hedy Lamarr (Hedwig Eva Maria Kiesler).

7.3 Adaptive Equalization

There are a variety of approaches to compensate for dispersive channels and their creation of ISI. These approaches attempt to mitigate the dispersive effects by applying filters that compensate for the channel. One approach, denoted zero-forcing, attempts to invert the spectral effect of the filter. Another approach is to employ an adaptive filter that minimizes the mean squared error between the signal at the output of the filter and the transmitted symbols [11]. Typically, training data is used to estimate the coefficients of these filters. The filters are then applied to the received signal for the data portion of the transmission.

In the frequency domain, we model the complex baseband received signal $Z(f)$ with

$$Z(f) = H(f)S(f) + N(f), \tag{7.12}$$

where $H(f)$, $S(f)$, and $N(f)$ are the frequency domain representations of the channel, transmitted baseband signal, and noise, respectively. If $S(f)$ has finite spectral support with some total bandwidth B at complex baseband, we assume that the signal at the receiver is filtered to this bandwidth, and thus, we only need to represent the portion of $H(f)$ accurately within B. By invoking Nyquist, we can accurately represent the signal and the portion of the channel within the signal bandwidth by using a sampled representation such that $\{\underline{s}\}_m = s(mT)$ and $\{\underline{h}\}_m = h(mT)$, where $T = 1/B$. Here, $s(t)$ and $h(t)$ are the time-domain representations of the transmitted signal and channel, respectively. Similarly, any filtering function $W(f) \Leftrightarrow w(t)$ can employ a sampled representation $\{\underline{w}\}_m = w(mT)$.

We often find it useful to represent the convolution with a vector matrix multiply, such that

$$\underline{h} * \underline{s} = \underline{h}\,\mathbf{S}. \tag{7.13}$$

Specifically, for the convolution of the sampled channel and transmitted signal, each row of the matrix $\mathbf{S}$ contains a different delayed version of $\underline{s}$, which we denote $\underline{s}_\tau$, given by

$$\mathbf{S} = \begin{pmatrix} \underline{s}_{\tau_1} \\ \underline{s}_{\tau_2} \\ \vdots \\ \underline{s}_{\tau_N} \end{pmatrix}. \tag{7.14}$$

More precisely, the mth row of the matrix $\mathbf{S}$ corresponds to the appropriate delayed signal of $\underline{s}$ that will be multiplied by the mth column of $\underline{h}$.

A common approach to estimate the sampled channel $\underline{\hat{h}}$ is given by the least-squared-error estimator, such that

$$\underline{\hat{h}} = \underline{z}\,\mathbf{S}^H\,(\mathbf{S}\,\mathbf{S}^H)^{-1}. \tag{7.15}$$

Here, we assume that $\underline{z}$ and corresponding $\mathbf{S}$ are taken from a region of the signal that contains training data.

7.3.1   Zero-Forcing Filter

The zero-forcing approach has the advantage of being intuitive, although it is suboptimal. The zero-forcing receiver applies an inverse filter $W(f) = 1/S^*(f)$ to the received signal such that

$$\hat{S}(f) = W^*(f)\,Z(f) \tag{7.16}$$
$$= W^*(f)\,H(f)\,S(f) + W^*(f)\,N(f). \tag{7.17}$$

This filter has the positive effect of flattening the system's spectral response. We refer to this approach as zero-forcing because in compensating for the spectral shape of the channel,

$$\underline{h} = (\cdots \quad h_{-m} \quad h_{-m+1} \quad \cdots \quad h_0 \quad \cdots \quad h_{m-1} \quad h_m \quad \cdots), \tag{7.18}$$

we zero all but one of the entries in the filtered channel in the idealized case,

$$\underline{w}^* * \underline{h} \propto (0 \quad \cdots \quad 0 \ 1 \ 0 \quad \cdots \quad 0). \tag{7.19}$$

However, for portions of the spectrum where the channel is small or even zero, this inverse has the effect of amplifying noise.

If we assume that the sampled channel $\underline{\mathbf{h}}$ can be approximated reasonably accurately with some finite set length, then we can implement the zero-forcing filter by

$$\underline{\mathbf{w}}^* * \underline{\mathbf{h}} = \underline{\mathbf{e}}_0 = (0 \quad \cdots \quad 0 \ 1 \ 0 \quad \cdots \quad 0), \tag{7.20}$$

where $\underline{\mathbf{e}}_0$ is a row vector of 0s with a 1 at its center. By representing the convolution with a vector matrix multiply, we have,

$$\underline{\mathbf{w}}^* \mathbf{H} = \underline{\mathbf{e}}_0 \tag{7.21}$$

$$\mathbf{H} = \begin{pmatrix} \underline{\mathbf{h}}_{\tau_1} \\ \underline{\mathbf{h}}_{\tau_2} \\ \vdots \\ \underline{\mathbf{h}}_{\tau_N} \end{pmatrix}, \tag{7.22}$$

where $\underline{\mathbf{h}}_{\tau_1}$ indicates a sampled channel $\underline{\mathbf{h}}$ that is shifted in time. We assume that the number of columns of $\mathbf{H}$ is equal to, or larger than, the number of rows. We can solve for $\underline{\mathbf{w}}$ by using the pseudoinverse, such that

$$\underline{\mathbf{w}}^* \mathbf{H} = \underline{\mathbf{e}}_0$$
$$\underline{\mathbf{w}}^* \mathbf{H} \mathbf{H}^H = \underline{\mathbf{e}}_0 \mathbf{H}^H$$
$$\underline{\mathbf{w}}^* = \underline{\mathbf{e}}_0 \mathbf{H}^H (\mathbf{H} \mathbf{H}^H)^{-1}. \tag{7.23}$$

Consequently, the zero-forcing equalizer applied to the data gives

$$\hat{\underline{\mathbf{s}}} = \underline{\mathbf{w}}^* * \underline{\mathbf{z}}. \tag{7.24}$$

We use Equation (7.15) to provide an estimate of the channel.

7.3.2 Wiener Filter

In order to approximately reconstruct the original signal with minimum error, a rake receiver with coefficients w_m is applied to the received signal:

$$\hat{s}(t) = \sum_m w_m^* z(t - mT), \tag{7.25}$$

where $\hat{s}(t)$ is the estimate of the transmitted signal and T is the sample period for which it is assumed that the sampling is sufficient to satisfy Nyquist sampling requirements. For an error $\epsilon(t)$, the mean squared error is given by

$$\epsilon(t) = \sum_m w_m^* z(t - mT) - s(t)$$

$$E\{\|\epsilon(t)\|^2\} = E\left\{ \left\| \sum_m w_m^* z(t - mT) - s(t) \right\|^2 \right\}. \tag{7.26}$$

The minimum mean squared error (MMSE) solution is found by taking a derivative of the mean squared error with respect to some parameter α of the filter
coefficient w_m:

$$\frac{\partial}{\partial \alpha} E\{\|\epsilon(t)\|^2\} = 0 \tag{7.27}$$

$$= \frac{\partial}{\partial \alpha} E\left\{ \left(\sum_m w_m^* z(t - m T) - s(t) \right) \left(\sum_n w_n^* z(t - n T_s) - s(t) \right)^* \right\}$$

$$= \frac{\partial}{\partial \alpha} E\left\{ \sum_{m,n} w_m^* z(t - m T_s) z^*(t - n T_s) w_n \right.$$

$$\left. - w_m^* z(t - m T_s) s^*(t) - s(t) z^*(t - n T_s) w_n + s(t) s^*(t) \right\}$$

$$= \frac{\partial}{\partial \alpha} \left(\sum_{m,n} w_m^* E\{z(t - m T_s) z^*(t - n T_s)\} w_n \right.$$

$$- w_m^* E\{z(t - m T_s) s^*(t)\} - E\{s(t) z^*(t - n T_s)\} w_m$$

$$\left. + E\{s(t) s^*(t)\} \right).$$

Constructing the filter vector $\mathbf{w}$, autocorrelation matrix $\mathbf{Q}$, and the cross-correlation
vector $\mathbf{v}$ by using the definitions

$$\{\mathbf{w}\}_m = w_m$$
$$\{\mathbf{Q}\}_{m,n} = E\{z(t - m T_s) z^*(t - n T_s)\}$$
$$\{\mathbf{v}\}_m = E\{z(t - m T_s) s^*(t)\}, \tag{7.28}$$

the derivative of the mean squared error or average error power[1] can be written as

$$\frac{\partial}{\partial \alpha} E\{\|\epsilon(t)\|^2\} = \frac{\partial}{\partial \alpha} \left(\mathbf{w}^H \mathbf{Q} \mathbf{w} - \mathbf{w}^H \mathbf{v} - \mathbf{v}^H \mathbf{w} + E\{s(t) s^*(t)\} \right)$$

$$= \left(\frac{\partial}{\partial \alpha} \mathbf{w}^H \right) [\mathbf{Q} \mathbf{w} - \mathbf{v}] + h.c., \tag{7.29}$$

where $h.c.$ indicates the Hermitian conjugate of the first term. This equation is solved
by setting the non-varying term to zero so that the filter vector $\mathbf{w}$ is given by

$$\mathbf{Q} \mathbf{w} = \mathbf{v}$$
$$\mathbf{w} = \mathbf{Q}^{-1} \mathbf{v}; \, \mathbf{Q} > 0. \tag{7.30}$$

This result is known as the Wiener–Hopf equation [11, 12]. The result can be formulated in more general terms, but this approach is relatively intuitive. Thus, the MMSE
estimate of the transmitted signal $\hat{s}(t)$ is given by

---

[1] Strictly speaking, the output of the filter should be parameterized in terms of energy per symbol, but it
is common to refer to this parameterization in terms of power. Because the duration of a symbol is
known, the translation between energy per symbol and power is a known constant.

$$\hat{s}(t) = \sum_m w_m^* \, z(t - m \, T_s)$$

$$w_m = \{\mathbf{Q}^{-1} \, \mathbf{v}\}_m \, . \tag{7.31}$$

7.4 Finite-Sample Weiner Formulation

Under the assumption that we have a finite number of samples of the received signal, we derive the adaptive equalizer. The estimated signal $\hat{\underline{t}} \in \mathbb{C}^{1 \times n_s}$ is given by

$$\hat{\underline{t}} = \mathbf{w}^H \tilde{\mathbf{Z}} \, , \tag{7.32}$$

where the adaptive equalizer is given by $\mathbf{w} \in \mathbb{C}^{M \times 1}$, and the received signal under the assumption of multiple possible delays is given by

$$\tilde{\mathbf{Z}} = \begin{pmatrix} \underline{\mathbf{z}}_{\tau_1} \\ \underline{\mathbf{z}}_{\tau_2} \\ \vdots \\ \underline{\mathbf{z}}_{\tau_M} \end{pmatrix} , \tag{7.33}$$

such that $\underline{\mathbf{z}}_{\tau} \in \mathbb{C}^{1 \times n_s}$ is the received signal at some delay τ. Note that the real channel spread, with N taps, is not the same as the number of delay distortions used by the adaptive filter: $M \neq N$. Finding the value of $\mathbf{w} \in \mathbb{C}^{M \times 1}$ minimizes the mean squared error, and shows that it is constructed from $\tilde{\mathbf{Z}} \in \mathbb{C}^{M \times 1}$ and $\underline{\mathbf{t}} \in \mathbb{C}^{1 \times n_s}$. We assume the model for the received signal $\underline{\mathbf{z}} \in \mathbb{C}^{1 \times n_s}$ is given by

$$\underline{\mathbf{z}} = \underline{\mathbf{h}} \tilde{\mathbf{T}} + \underline{\mathbf{n}} \, , \tag{7.34}$$

where $\underline{\mathbf{n}} \in \mathbb{C}^{1 \times n_s}$ indicates n_s independent samples of zero-mean circularly symmetric complex Gaussian noise. The N tap dispersive channel is represented by

$$\underline{\mathbf{h}} = (h_{\tau_1} \quad h_{\tau_2} \quad \cdots \quad h_{\tau_N}) \, , \tag{7.35}$$

and the training signal matrix $\tilde{\mathbf{T}} \in \mathbb{C}^{N \times n_s}$ with N delay distortions of the training sequence $\underline{\mathbf{t}} \in \mathbb{C}^{1 \times n_s}$ with delay τ is given by

$$\tilde{\mathbf{T}} = \begin{pmatrix} \underline{\mathbf{t}}_{\tau_1} \\ \underline{\mathbf{t}}_{\tau_2} \\ \vdots \\ \underline{\mathbf{t}}_{\tau_N} \end{pmatrix} . \tag{7.36}$$

The error vector $\underline{\boldsymbol{\varepsilon}} \in \mathbb{C}^{1 \times n_s}$ is given by

$$\underline{\boldsymbol{\varepsilon}} = \hat{\underline{\mathbf{t}}} - \underline{\mathbf{t}} = \mathbf{w}^H \tilde{\mathbf{Z}} - \underline{\mathbf{t}} \tag{7.37}$$

so that an estimate of the mean squared error is given by $\|\underline{\boldsymbol{\varepsilon}}\|^2 / n_s$.

To perform the optimization, we evaluate the derivative with respect to the real parameter α under the assumption that the filter coefficients are an undetermined function of this parameter $\mathbf{w}(\alpha)$:

$$0 = \frac{\partial}{\partial\alpha}\frac{\|\underline{\varepsilon}\|^2}{n_s}$$

$$= \frac{\partial}{\partial\alpha}\left([\mathbf{w}^H\tilde{\mathbf{Z}} - \underline{\mathbf{t}}][\mathbf{w}^H\tilde{\mathbf{Z}} - \underline{\mathbf{t}}]^H\right)$$

$$= \frac{\partial}{\partial\alpha}\left([\mathbf{w}^H\tilde{\mathbf{Z}} - \underline{\mathbf{t}}][\tilde{\mathbf{Z}}^H\mathbf{w} - \underline{\mathbf{t}}^H]\right)$$

$$= \left(\frac{\partial}{\partial\alpha}\mathbf{w}^H\right)\left(\tilde{\mathbf{Z}}[\tilde{\mathbf{Z}}^H\mathbf{w} - \underline{\mathbf{t}}^H]\right) + h.c. \tag{7.38}$$

The entire expression is zero if

$$0 = \tilde{\mathbf{Z}}[\tilde{\mathbf{Z}}^H\mathbf{w} - \underline{\mathbf{t}}^H]$$

$$\tilde{\mathbf{Z}}\tilde{\mathbf{Z}}^H\mathbf{w} = \tilde{\mathbf{Z}}\underline{\mathbf{t}}^H$$

$$\mathbf{w} = (\tilde{\mathbf{Z}}\tilde{\mathbf{Z}}^H)^{-1}\tilde{\mathbf{Z}}\underline{\mathbf{t}}^H. \tag{7.39}$$

The adaptive equalizer can then be applied to portions of the data that are not associated with the training data.

7.4.1   Weiner Filter Example

If we transmit with a quadrature-phase-shift keying (QPSK) signal, the constellation represented at baseband is shown in Figure 7.4. We have a flat-fading channel and in the presence of circularly symmetric complex additive Gaussian noise. For this example, we assume a received SNR of 20 dB and have 200 training samples. The received constellation, after the channel compensation for attenuation and phase rotation, is presented in Figure 7.5.

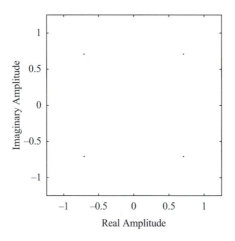

Figure 7.4 Transmitted QPSK complex baseband constellation.

Even in the presence of noise, the constellation points are easily distinguishable, as one would expect at this SNR.

As an example, we consider the sampled dispersive channel that is defined by the vector $\underline{\mathbf{h}}$:

$$\underline{\mathbf{h}} \propto (0.1 \quad -0.3\,i \quad 1 \quad 0.3\,i \quad 0.2 - 0.7\,i \quad 0.1\,i) \tag{7.40}$$

$$\|\underline{\mathbf{h}}\| = 1\,. \tag{7.41}$$

We can translate the temporal representation of the channel to the spectral shaping of the channel. The relative effect of the received energy spectral density is depicted in Figure 7.6. We represent the spectral density as a function of normalized frequency. To convert to the actual frequency, one would multiply this frequency by the complex

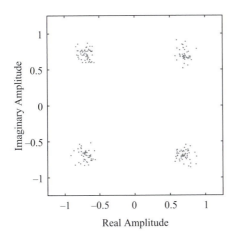

Figure 7.5 200 example received QPSK complex baseband constellation points under the assumption of a flat-fading channel with an SNR of 20 dB.

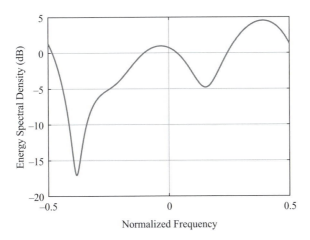

Figure 7.6 Effect of the spectral shaping introduced by the example dispersive channel.

sample rate. In the figure, the frequency-selective aspects of the channel become clear. There is over a 20 dB variation as a function of frequency.

Because of the delay spread of the dispersive channel, ISI is introduced. The addition of both ISI and noise produces the unrecognizable received constellation depicted in Figure 7.7. Because of the ISI, we moved from an easily decodable received constellation, represented in Figure 7.5, to this unrecognizable one.

To compensate for the channel effects, we implement the finite-sample MMSE adaptive filter presented in Equation (7.39). Implicit in this discussion is that the data is synchronized so that the delays of the received signal fall well within the extent of the adaptive filter **w**. For this example, we select an adaptive filter length of 25, which is comfortably larger than the extent of the delay spread. We present the real and imaginary components of the constructed filter in Figure 7.8. Typically, adapted

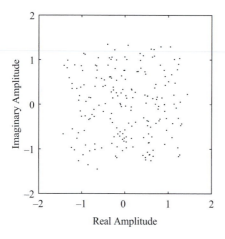

Figure 7.7 200 example received QPSK complex baseband constellation points under the assumption of an example frequency-selective channel $\underline{\mathbf{h}}$ with an SNR of 20 dB.

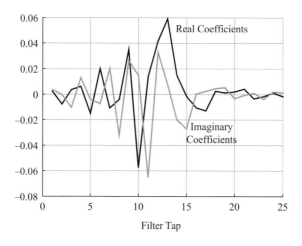

Figure 7.8 Real and imaginary coefficients of the adaptive filter as a function of sample delay.

filters have larger values toward the center taps and typically fall in magnitude as the delays move away from the center. If the filter is not sufficiently long, if the signal is not properly centered in the filter delay, or if there is insufficient training data, the filter coefficients will not have this shape. We display the effective spectral shape of the adaptive filter in Figure 7.9. The shape approximately compensates for the channel spectral shape observed in Figure 7.6. The product of these two spectra is given in Figure 7.10. The product spectrum is approximately flat. It is worth noting that the adaptive filter does not exactly compensate for the channel shape. There are multiple reasons for this lack of perfect compensation. When the adaptive filter magnifies a weak region of the spectrum, it effectively amplifies the noise. Another reason is that it, in general, takes an infinite number of taps for a finite-impulse-response (FIR) filter to invert another FIR filter.

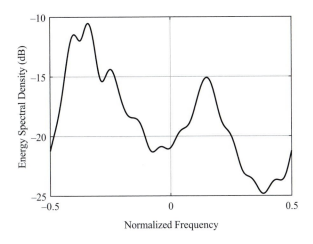

Figure 7.9 Energy spectral density shaping of the adaptive filter.

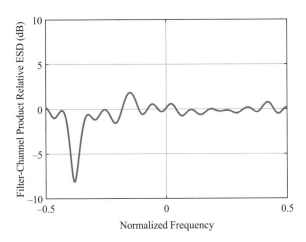

Figure 7.10 Energy spectral density of the channel after adaptive filtering.

We depict the resulting constellation observed in the receiver after the adaptive filter in Figure 7.11. The observed points in the receive constellation are well separated, although the effective SNR is not as high as the channel without dispersion.

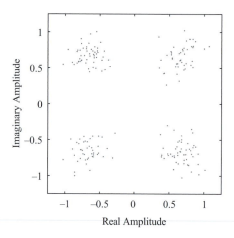

Figure 7.11 200 example received QPSK complex baseband constellation points after adaptive fitering under the assumption of an example frequency-selective channel $\underline{\mathbf{h}}$ with an SNR of 20 dB.

7.5   OFDM Modulation

Orthogonal frequency-division multiplexing (OFDM) [13] takes advantage of two observations. First, narrowband signals cannot resolve delay spread. Consequently, a flat-fading model can be applied in each subcarrier. Second, implementing fast Fourier transforms (FFTs) is relatively inexpensive, computationally.

If we consider a modulation that is constrained to the bandwidth B, then the Nyquist period of time is given by $T = 1/B$. We construct a narrowband signal that is largely (but not completely) contained within an approximate bandwidth of $B' = B/N$. This symbol would have a duration of at least NT. If the narrowband signal had a top hat pulse shape, the spectral content would be proportional to a sinc function, which is given by

$$
\begin{aligned}
\int_{-NT/2}^{NT/2} dt\, a\, e^{-i2\pi tf} &= a\, \frac{e^{-i2\pi tf}}{-i2\pi f}\bigg|_{-NT/2}^{NT/2} \\
&= a\left(\frac{e^{-i2\pi NT/2f}}{-i2\pi f} - \frac{e^{+i2\pi NT/2f}}{-i2\pi f}\right) \\
&= a\, \frac{\sin(\pi NT f)}{\pi f} \\
&= a\, NT\, \mathrm{sinc}(NT f).
\end{aligned} \tag{7.42}
$$

We display an example for $NT = 16$ in Figure 7.12.

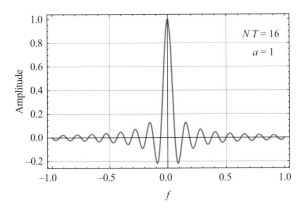

Figure 7.12 Spectral content of a top hat of duration $NT = 16$.

Because most of the band is relatively empty, we can imagine adding another narrowband signal to be transmitted at the same time. It would be reasonable to place the second sinc function at the spectral zero of the first sinc. This zero occurs when the argument of the sine function is zero with the exception of zero frequency, as can be observed in

$$\text{sinc}(NT f_{\text{off}}) = \frac{\sin(\pi \, NT f_{\text{off}})}{\pi \, NT f_{\text{off}}}$$

$$\Rightarrow \sin(\pi \, NT f_{\text{off}}) = 0 \, ; \qquad f_{\text{off}} \neq 0$$

$$\Rightarrow \pi \, NT f_{\text{off}} = m \, \pi \, ; \qquad m \in \mathbb{Z}, m \neq 0$$

$$f_{\text{off}} = \frac{m}{NT} . \tag{7.43}$$

The closest spacing can be found when $m = 1$, which is depicted in Figure 7.13. We can extend this sequence of subcarriers to consider a larger number of subcarriers. In Figure 7.14, we consider the combination of 16 subcarriers. In this case, each subcarrier has an independent BPSK modulation for a single OFDM symbol.

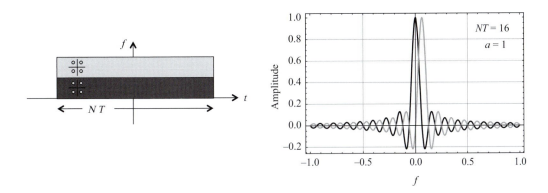

Figure 7.13 Two spectral sinc functions separated at critical spacing.

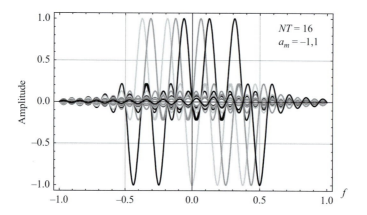

Figure 7.14 16 sinc functions representing OFDM modulations with BPSK in each subcarrier $(-1, +1, +1, -1, +1, +1, +1, -1, +1, +1, -1, -1, +1, -1, -1, -1)$.

Section 15.4.2 tells us that to shift a signal by a frequency $m/(NT)$, we multiply the original time-domain function by a phase ramp defined by

$$\tilde{s}(t) = e^{i 2\pi\, t m/(NT)}\, s(t)$$

$$\tilde{s}(t) = e^{i 2\pi\, t m/(NT)}\, a_m,\tag{7.44}$$

where the subcarrier channel has the constant value a_m. If we assume units so that $T = 1$ and consider critical sampling, we can replace t with n so that

$$\tilde{s}(n) = e^{i 2\pi\, n m/N}\, a_m.\tag{7.45}$$

If we fill all N available subcarriers, and note that the complex baseband output at a given time will be the sum across subcarriers, we find that the output signal for the nth sample x_n is given by

$$x_n = \frac{1}{N} \sum_m e^{i 2\pi\, n m/N}\, a_m,\tag{7.46}$$

where we have added the $1/N$ normalization for convenience. We observed that this form is simply the inverse discrete Fourier transform (DFT) from Section 15.7. In practical implementations, this transformation is typically evaluated by using an inverse FFT (IFFT). For a set of symbols assigned to various frequencies $\mathbf{a}$, the output is given by $\mathbf{x}$:

$$\mathbf{x} = \mathbf{F}^{-1}\, \mathbf{a}\tag{7.47}$$

$$\mathbf{F}^{-1} = \frac{1}{N}\mathbf{F}^H$$

$$= \frac{1}{N}\begin{pmatrix} \beta^{0\cdot 0} & \beta^{0\cdot 1} & \beta^{0\cdot 2} & \cdots & \beta^{0\cdot(N-1)} \\ \beta^{1\cdot 0} & \beta^{1\cdot 1} & \beta^{1\cdot 2} & \cdots & \beta^{1\cdot(N-1)} \\ \vdots & & & \ddots & \\ \beta^{(N-1)\cdot 0} & \beta^{(N-1)\cdot 1} & \beta^{(N-1)\cdot 2} & \cdots & \beta^{(N-1)\cdot(N-1)} \end{pmatrix},\tag{7.48}$$

where $\beta = \alpha^* = e^{i\,2\pi/N}$. To reverse the process and recover estimates for each subcarrier, we perform a corresponding FFT in the receiver. Note that the channel will cause scaling and phase rotations that are frequency-dependent in general, as observed in Figure 7.15.

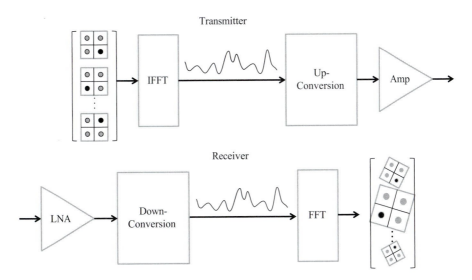

Figure 7.15 Block diagram for a transmitter and a receiver that employs OFDM.

As a practical matter, the subcarriers near the edges of the spectrum are typically not used. These zeroed subcarriers help to assure spectral confinement.

7.5.1   Subcarrier Spacing

Determining the number of subcarriers is an engineering trade and depends upon a number of factors. Optimization is a function of numerous parameters, including the total signal bandwidth, the delay spread, the spectral efficiency of the modulation and error correction coding, the performance of the error correction coding, the computational cost, and the stability of the channel. In the spectral domain, one can view this requirement by considering whether the channel can be accurately modeled by a flat-fading channel within a subcarrier, as depicted in Figure 7.16. In this domain, one can see that the spectral coherence interval (how quickly the channel changes as a function of frequency) must be large compared to the subcarrier spacing. This directly suggests that the number of subcarriers is at least as large as required to satisfy this discussion. Because IFFTs and FFTs are particularly efficient for powers of 2, one often rounds up to the nearest power of two.

An alternative view of the same discussion is provided by considering the length of the OFDM symbol in time compared to the delay spread of the channel. The signal observed at the receiver is given by the convolution of the channel impulse response and the transmitted signal. If the length of the channel in time is a reasonable portion

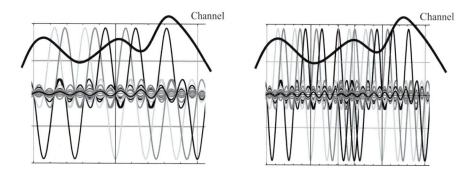

Figure 7.16 Frequency-domain representations of frequency-dependent channels to suggest number of subcarriers.

of the length of the OFDM symbol, then the symbol within a subcarrier will suffer significant distortion. However, if the delay spread is short compared to the OFDM length, then the distortion will be minimal, as observed in Figure 7.17. As the number of subcarriers increases, the temporal duration of the OFDM symbol increases. The fundamental length of the OFDM symbol is the ratio of the number of subcarriers to the total bandwidth of the signal. A not unreasonable rule of thumb is that the OFDM symbol must be several times longer than the delay spread. One might wonder, why do we not simply make the number of subcarriers arbitrarily large? As the duration of the OFDM symbol becomes large, the assumption of a static channel over the symbol comes into question. At some point, channel dynamics causes more distortion than frequency selectivity. As a secondary issue, the computation cost increases with the number of subcarriers. Finally, if the subcarrier spacing is too small, local oscillator errors or Doppler-frequency offsets can cause subcarrier misalignment between the transmitter and receiver. This misalignment causes inter-subcarrier interference (ICI).

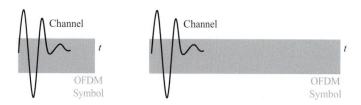

Figure 7.17 Time-domain representations of frequency-dependent channels to suggest number of subcarriers.

7.5.2   Cyclic Prefix

Because of delay spread induced by the channel, we have an additional issue with which we must contend. As we depict in Figure 7.18, the multiple resolvable delays of the dispersive channel are convolved with the OFDM symbol. The energy of the

OFDM symbol observed at the receiver is spread over a longer period of time than the transmitted OFDM symbol. The FFT contained within the receiver is performed for N critical samples. The block of signal on which the FFT is applied will miss some of the OFDM symbol for all but just the right delay. The contribution of a mismatched delay is smaller than the intended duration of NT. Consequently, its spectral shape is wider because the spectral width is inversely related to the observed signal's duration (Figure 7.19).

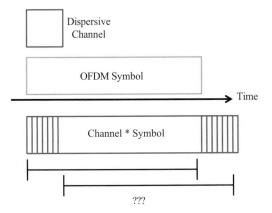

Figure 7.18 How delay spread breaks inter-subcarrier orthogonality.

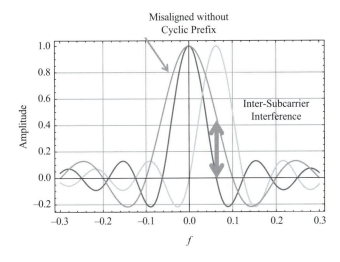

Figure 7.19 Mismatched received duration breaks inter-subcarrier orthogonality by broadening sinc of subcarriers.

To compensate for this effect, we exploit the observation that circular shifts in the input vector of the DFT introduce frequency-dependent phase rotations but otherwise do not affect the output. Because the channel introduces phase rotations, the overall frequency-dependent effects of the phase shifts need to be estimated and compensated. If we copy a portion of the beginning of the OFDM symbol and append it to the end of the symbol (as we depict in Figure 7.20), or if we copy a bit of the end of the OFDM symbol and prepend it to the beginning of the symbol, then we can produce a received OFDM symbol for which delays that fall off the end show up at the beginning. Thus, orthogonality between subcarriers is recovered. Here, we have assumed that the cyclic prefix is sufficiently long to effectively contain the delay spread.

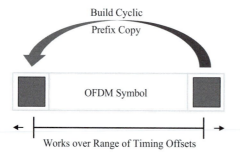

Figure 7.20 Adding a cyclic prefix to recover inter-subcarrier orthogonality.

7.5.3   OFDM Channel Estimation

The density of subcarriers is typically selected so that, within each subcarrier, a narrowband channel approximation can be employed. Each subcarrier has a different scaling and phase. In order to interpret the transmitted signal, the propagation channel for each subcarrier must be compensated separately. While there are numerous techniques, in general they employ training data that sparsely samples the channel across the frequency domain by using only a subset of the subcarriers for training. Subcarrier channels between these estimates are determined by some sort of interpolation. As an example, the data subcarrier channel coefficients are determined via linear interpolation, as we depict in Figure 7.21. In this case, this frequency-selective channel overwhelms linear interpolation, so subcarrier density and interpolation technique are important.

7.6   Dispersive Channel Model

While the specifics of each physical environment are different, there are some common models that are useful for simulating representative dispersive channels. To help in simulations of communications systems, we often separate the channel effects into a component that incorporates the typical attenuation (which we discuss in Section 7.6.1) and another component that incorporates the fluctuating effects of scatterers. On a linear scale, these models are multiplied. An example model is given by the

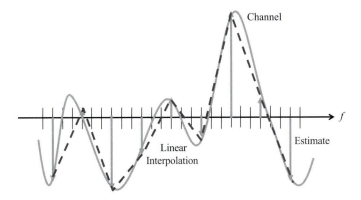

Figure 7.21 Interpolate sparse channel estimates across subcarriers.

exponential envelope that addresses the fluctuating effect of many scatters. The motivation for this model is that the earliest arriving multipath signals take fewer bounces, so they are stronger on average, although any given channel is random. If the exponential is aligned with the sampling lattice (which would not be generally expected), then an example channel vector $\underline{h}$ would be given by

$$\underline{h} \propto (e^{-0T_s/\tau} \, g_0 \quad e^{-1T_s/\tau} \, g_1 \quad e^{-2T_s/\tau} \, g_2 \cdots) \tag{7.49}$$

$$\underline{h} = a \, \frac{(e^{-0T_s/\tau} \, g_0 \quad e^{-1T_s/\tau} \, g_1 \quad e^{-2T_s/\tau} \, g_2 \cdots)}{\sqrt{\dfrac{1}{1-e^{-2T_s/\tau}}}}, \tag{7.50}$$

where τ defines the shape parameter, T_s is the sample interval, a is the average amplitude attenuation, and g_m is independently drawn from a zero-mean circularly symmetric complex Gaussian distribution, as described in Section 14.1.6. The received signal would be given by $\underline{z} = \underline{h} * \underline{s} + \underline{n}$. The power-weighted mean delay is given by

$$\tau_0 = \frac{E\left\{ \sum_{m=0}^{\infty} mT_s \, \|\{\underline{h}\}_m\|^2 \right\}}{E\left\{ \sum_{m=0}^{\infty} \|\{\underline{h}\}_m\|^2 \right\}} \tag{7.51}$$

$$= \frac{\sum_{m=0}^{\infty} mT_s \, e^{-2mT_s/\tau}}{\sum_{m=0}^{\infty} e^{-2mT_s/\tau}} \tag{7.52}$$

$$= \frac{T_s}{e^{2T_s/\tau} - 1}. \tag{7.53}$$

The root mean squared (RMS) delay spread τ_{RMS} (where we have shifted the axis so that the mean delay is zero) is given by

$$\tau_{RMS} = \sqrt{\frac{E\left\{ \sum_{m=0}^{\infty} (mT_s - \tau_0)^2 \, \|\{\underline{h}\}_m\|^2 \right\}}{E\left\{ \sum_{m=0}^{\infty} \|\{\underline{h}\}_m\|^2 \right\}}} \tag{7.54}$$

$$= \frac{T_s \, e^{T_s/\tau}}{e^{2T_s/\tau} - 1}. \tag{7.55}$$

As an example, we consider a scenario in which the bandwidth is 10 MHz, so that the complex Nyquist sample interval is 100 ns. If the channel amplitude typically falls like τ = 200 ns, then the RMS delay spread is given by τ_{RMS} = 96 ns. We display an example of four drawn channels in Figure 7.22. The corresponding power spectral densities as a function of normalized frequency are given in Figure 7.23.

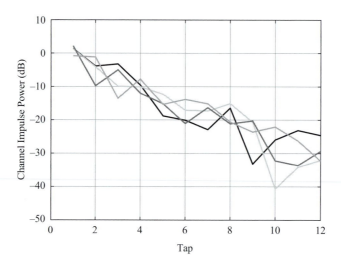

Figure 7.22 Examples of impulse responses of four independently drawn exponential channels.

7.6.1   Obstructive Channel Attenuation

In complicated scattering environments, it is not reasonable to assume a line-of-sight model (discussed in Section 2.2.2) would be valid. It is very difficult to actually calculate what sort of attenuation would be expected. A full electromagnetic model might work, but we rarely know all the materials and geometries sufficiently well to create it. Modified ray-tracing models are commonly used to aid base station placement analysis, but the actual attenuation is only approximated by these models.

Another class of model is phenomenological, which is used to provide typical overall attenuation. For a given environment, a number of attenuation measurements are made, and a typical attenuation is determined as a function of a few environmental parameters. For simulations, we often use these models to provide typical overall attenuation and then assume a multiplicative fading model extension.

For cellular band wireless attenuation, the Okumura–Hata model [14] is a commonly used model. The model fits attenuation empirical data between a handset and a base station. For an urban environment, on a decibel scale, the attenuation approximation L_U (dB) is given by

$$L_U = 69.55 + 26.16 \log_{10}(f) - 13.82 \log_{10}(h_B) - C_H$$
$$+ [44.9 - 6.55 \log_{10}(h_B)] \log_{10}(d), \qquad (7.56)$$

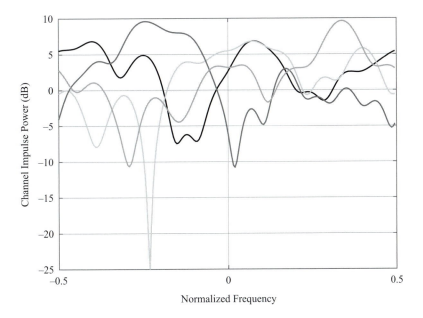

Figure 7.23 Examples of power spectral efficiency of four independently drawn exponential channels as a function of normalized frequency.

where h_B is the height of the base station in meters; h_M is the height of the mobile station (handset) in meters; f is the carrier frequency in megahertz (MHz); C_H is the antenna height correction factor; and d is the distance between mobile and base stations.

For smaller cities, the antenna height correction factor C_H is given by

$$C_H = 0.8 + [1.1\ \log_{10}(f)]\,h_M - 1.56\ \log_{10}(f). \tag{7.57}$$

For larger cities it is given by

$$C_H = \begin{cases} 8.29\,[\log_{10}(1.54\,h_M)]^2 - 1.1 & ;\quad 150 \le f \le 200 \\ 3.2\,[\log_{10}(11.57\,h_M)]^2 - 4.97 & ;\quad 200 \le f \le 1500 \end{cases}. \tag{7.58}$$

This is a very specific model with a limited range of applicability. While it is of some use, we show it primarily as an example of phenomenological models.

Problems

7.1 For the channel impulse response given by $\underline{h} = [1 \quad 0.3\,i \quad 0.2 - 0.1\,i]$, evaluate the shape of the power spectral density of the channel as a function of frequency by evaluating the autocorrelation.

7.2 While channel impulses are determined by the physical propagation environment, we often describe them as random vectors of complex coefficients for a lattice of delays. A common, although not perfect, model for delay spread is given by channel coefficients sampled from independent zero-mean unit-variance complex Gaussians and weighted by an exponential, such that

$$h_m = a e^{-km} g_m, \quad \text{for} \quad m \geq 0,$$

and $h_m = 0$ otherwise, where a is the channel attenuation, k defines the distribution shape, and the random variables g_m are drawn from the Gaussian distribution. By assuming that the complex channel samples are provided at an interval of T, find the expected average delay and variance of the delay spread of the channel as a function of k. Hint: use the expected value of $\|h_m\|^2$ to weight the delays.

7.3 By using the form of the channel defined in Problem 7.2, plot 100 random QPSK constellation points and the magnitude squared of the FFT (and use zero-padding to smooth the shape) of the channel for a single draw for a given channel defined by
(a) no delay spread
(b) $k = 0.3$
(c) $k = 1$
(d) $k = 3$
(e) $k = 10$
(f) $k = 25$

7.4 Derive the adaptive equalizer so that the estimated signal $\hat{\mathbf{t}} \in \mathbb{C}^{1 \times n_s}$ is given by

$$\hat{\mathbf{t}} = \mathbf{w}^H \tilde{\mathbf{Z}},$$

where the adaptive equalizer is given by $\mathbf{w} \in \mathbb{C}^{M \times 1}$ and the received signal under the assumption of multiple possible delays is given by

$$\tilde{\mathbf{Z}} = \begin{pmatrix} \underline{\mathbf{z}}_{\tau_1} \\ \underline{\mathbf{z}}_{\tau_2} \\ \vdots \\ \underline{\mathbf{z}}_{\tau_M} \end{pmatrix},$$

such that $\underline{\mathbf{z}}_{\tau} \in \mathbb{C}^{1 \times n_s}$ is the received signal at some delay τ. Note that the real channel spread, with N taps, is not the same as the number of delay distortions used by the adaptive filter: $M \neq N$. To find the value of $\mathbf{w} \in \mathbb{C}^{M \times 1}$, minimize the mean squared error and show that it is constructed from $\tilde{\mathbf{Z}} \in \mathbb{C}^{M \times 1}$ and $\mathbf{t} \in \mathbb{C}^{1 \times n_s}$. To do this optimization review the $\partial \mathbf{w}^H(\alpha)/\partial \alpha$ approach discussed in Section 13.8. We assume the model for the received signal $\underline{\mathbf{z}} \in \mathbb{C}^{1 \times n_s}$ is given by

$$\underline{\mathbf{z}} = \underline{\mathbf{h}} \tilde{\mathbf{T}} + \underline{\mathbf{n}},$$

where $\underline{\mathbf{n}} \in \mathbb{C}^{1 \times n_s}$ indicates n_s independent samples of circularly symmetric complex Gaussian noise, the N tap dispersive channel is represented by

$$\underline{\mathbf{h}} = (h_{\tau_1} \quad h_{\tau_2} \quad \cdots \quad h_{\tau_N}),$$

and the training signal matrix $\tilde{\mathbf{T}} \in \mathbb{C}^{N \times n_s}$ with N delay distortions of the training sequence $\underline{\mathbf{t}} \in \mathbb{C}^{1 \times n_s}$ with delay τ is given by

$$\tilde{\mathbf{T}} = \begin{pmatrix} \underline{\mathbf{t}}_{\tau_1} \\ \underline{\mathbf{t}}_{\tau_2} \\ \vdots \\ \underline{\mathbf{t}}_{\tau_N} \end{pmatrix}.$$

The error vector $\underline{\varepsilon} \in \mathbb{C}^{1 \times n_s}$ is given by

$$\underline{\varepsilon} = \hat{\underline{\mathbf{t}}} - \underline{\mathbf{t}} = \mathbf{w}^H \tilde{\mathbf{Z}} - \underline{\mathbf{t}},$$

so that the mean squared error is given by $\|\underline{\varepsilon}\|^2 / n_s$.

7.5 Draw the block diagrams for radios that transmit and receive an OFDM modulation.

7.6 Estimate typical inter-carrier interference relative power for an OFDM symbol without a cyclic prefix whose reception is misaligned by 25 percent.

7.7 This is an approximate OFDM design problem. For a system with a 10 MHz bandwidth, (1) determine approximately the maximum subcarrier spacing, and (2) determine the number of subcarriers required to satisfy this maximum subcarrier bandwidth. (3) Assume that the number of subcarriers is a power of 2 such that 2^n ; $n \in \mathbb{Z}$. Assume that the maximum delay spread is given by
(a) indoor, 20 ns
(b) large building, 200 ns
(c) outside, 2 μs
(d) metal ship's hold, 200 μs

7.8 Consider an OFDM approach that employs eight subcarriers (which is admittedly too small to be all that useful).
(a) For a set of symbols given in vector **s**, what is a simple matrix-vector form for the transmit sequence in time denoted by column vector **x**? *Define any terms clearly.*
(b) What is the transmitted time-domain sequence for $\mathbf{s} = (1, 0, 0, 0, 0, 0, 0, 0)^T$?
(c) What is the transmitted time-domain sequence for $\mathbf{s} = (1, -1, 0, 0, 0, 0, 0, 0)^T$?

7.9 Consider the design of an OFDM modulated signal. Assume that the transmitter operates at a 10 MHz complex sample rate and that the maximum expected delay spread is less than 1 μs.
(a) What is the complex Nyquist sample period T_s (which is not the length of the OFDM symbol)?
(b) To assure orthogonality, what is the minimum number of samples to use for the cyclic prefix?
(c) If the OFDM symbol has 100 subcarriers, what is the subcarrier spacing in frequency?
(d) If the OFDM symbol has 100 subcarriers, for how long in time is the OFDM symbol including the cyclic prefix determined previously?

7.10 By using the representative channel defined by

$$\underline{h} = [0.1 \quad -0.3\,i \quad 1 \quad 0.3\,i \quad 0.2 - 0.7\,i \quad 0.1\,i],$$

design and build an OFDM approach that uses QPSK subcarrier modulation. Identify subcarrier spacing, training density, and cyclic prefix length. Show that the QPSK symbols can be reconstructed well in simulation while using this channel.

8   Error Correcting Codes

In this chapter, we discuss error correcting codes. We review the idea of hard versus soft decisions as types of symbol estimates that we provide to the decoder. We introduce the concepts of parity bits – which provide redundancy – and coding rate. We develop a simple, if flawed, toy systematic linear block code and use this code to demonstrate the concepts of generator matrix, hamming distance, and parity check matrix. By using the parity check matrix, we construct the syndrome and use this vector to perform error correction. To provide a set of viable systematic linear block codes, we introduce Hamming codes. We also introduce convolutional codes and relate the mathematical and shift-register block diagram forms. To enable decoding, we discuss the trellis diagram and use this diagram to motivate Viterbi decoding.

As we have discussed previously, error correcting codes are extremely important in communications systems. The encoding and decoding are essentially the first thing you do with the information bits in the transmitter and the last step to reconstruct them in the receiver, as seen in Figure 8.1. Error correcting codes are a practical way to approach the performance promised by the Shannon limit (discussed in Chapter 3).

Theoretically, one does not need error correcting codes as we describe them here. We could construct a dictionary of N pseudorandom complex baseband sequences of length M chips. We could then encode $\log_2(N)$ bits by selecting one of these sequences and transmitting it. The receiver would compare the received sequence to each of the possible sequences in the dictionary. The sequence that was closest to the received signal would be declared the correct sequence; thus, the original information bits could be determined. However, as the number of bits grows, this approach becomes computationally untenable. Imagine a coded block of 100 information bits, which is considered relatively short. This would correspond to 2^{100} or about 1.3×10^{30} test sequences at the receiver for each 100-bit block. Error correcting codes enable one to approach the performance produced by this type of coding with significantly reduced computational complexity.

In this chapter, we use block codes as our primary tool for introducing error correcting codes. However, we do provide a short introduction to convolutional codes in Section 8.8.

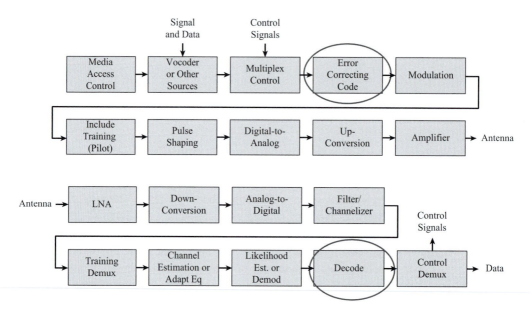

Figure 8.1 The block diagram of the transmitter and receiver chains with coding and decoding blocks identified.

8.1   Soft versus Hard Decoding

In the receiver, one can imagine a demodulation procedure that immediately converts a received amplitude to a bit decision. As an example, for binary-phase-shift keying (BPSK), displayed in Figure 8.2, if the real part of the channel-compensated received complex baseband signal is greater than 0, then it would be declared that the received signal was generated by a bit 1; otherwise, a bit 0 would be declared. These hard decisions would be passed on to the error correction code.

For a soft-decoder approach, the distance from the correct constellation point could be passed on to the decoder. Under the assumption of Gaussian noise, the square of the distance is proportional to the logarithm of the likelihood. For this BPSK case, if the signal has a large real component, then it is particularly unlikely to have been the result of a −1 point on the constellation; conversely, if the real component has a value near 0, then either ±1 are almost equally likely. Unsurprisingly, the use of likelihoods rather than hard decisions improves the performance of error correction codes. Because we intend to provide an introduction to error correcting codes, we will only consider the simpler presentation of hard decisions and corresponding hard decoders in this text.

8.2   Parity Bits

The fundamental tool for error correcting is to add redundant bits to the information bits. In general there is a complicated interconnection between a group of information bits and the corresponding parity bits. The simplest version of this information-to-parity-bit construction is the repetition code. Imagine each information bit of 0 or 1 was transmitted three times. Consequently, there are two parity bits. At the receiver, a simple voting circuit can be applied to the estimated bits to provide an improved estimate of the original information bit. This is a relatively poor code, but it is easy to understand. The code can correct for any single received estimated bit error out of the three bits in the encoded sequence.

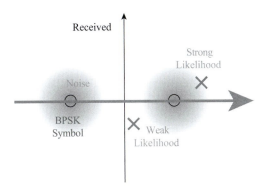

Figure 8.2 Examples of strong and weak likelihoods for a received BPSK constellation.

It is often useful to denote the coding rate. The coding rate r is given by the ratio of the number of information bits n_{info} to the number of total bits $n_{info} + n_{parity}$ in a coding block, as expressed by

$$r = \frac{n_{info}}{n_{info} + n_{parity}} \tag{8.1}$$

$$n_{total} = n_{info} + n_{parity}, \tag{8.2}$$

where n_{parity} is the number of parity bits in the coding block. For our simple redundancy code, there are two parity bits for each information bit; consequently, the coding rate is $1/(1+2) = 1/3$.

It is important to note that the spectral efficiency (b/s/Hz) bounded by the Shannon limit (discussed in Chapter 3) is associated with the information bit rate only. Parity bits are not counted because they only provide information that is redundant to the information bits. For a typical digital modulation, the spectral efficiency η is given by the product of the code rate r and the number of bits required to specify a constellation point. The spectral efficiency is given by

$$\eta = r\, n_{sym} \text{ (b/s/Hz)} \tag{8.3}$$

$$n_{sym} = \log_2(n_{const}), \tag{8.4}$$

where n_{const} is the number of points on the constellation and n_{sym} is the number of bits required to specify a given constellation point. As an explicit example, consider a quadrature-phase-shift keying (QPSK) constellation and a coding rate $r = 4/7$: The spectral efficiency is given by

$$n_{sym} = \log_2(4) = 2 \tag{8.5}$$

$$\eta = \frac{4}{7} 2 = \frac{8}{7}. \tag{8.6}$$

8.3   Simple Example Code

For pedagogical reasons we will now consider a bad toy block code. For our toy code, we use a block code suggested in Reference [2]. First, codes that are very short do not perform well. Second, this code is fundamentally flawed. However, the code has a number of interesting characteristics that are useful for introducing more capable codes.

If we have a linear binary block code, then code words can be generated by simply multiplying a row vector containing information bits and the generator matrix. This particular block code is systematic, which means information bits can be seen directly in the code words. In this case, the first two bits are the information bits. All code words $\underline{c} \in \mathbb{Z}_2^{1 \times n}$, which are represented here by row vectors of binary ($\mathbb{Z}_2$) elements with values of 1 or 0, are constructed from the equation

$$\underline{c} = \underline{x}G, \qquad G \in \mathbb{Z}_2^{k \times n}, \tag{8.7}$$

where a row vector of information bits is given by $\underline{x} \in \mathbb{Z}_2^{1 \times k}$ for the generator matrix G. One of the tricks in performing this calculation is to remember to perform the calculations by using the correct field – in this case, a binary or, equivalently, Galois field of size 2, denoted GF(2). For our toy linear systematic binary block code, the generator matrix is given by

$$G = \begin{pmatrix} 1 & 0 & | & 1 & 0 & 0 \\ 0 & 1 & | & 1 & 1 & 1 \end{pmatrix} \tag{8.8}$$

$$G = (I \quad P)$$

$$P = \begin{pmatrix} 1 & 0 & 0 \\ 1 & 1 & 1 \end{pmatrix}.$$

Because the code is systematic, we can partition the generator matrix into a $k \times k$ identity matrix I and a $k \times (n - k)$ parity-generating matrix P. The vertical line in Equation (8.8) has no mathematical meaning, and is only there to guide the eye by separating the identity matrix and the parity-generating matrix.

We can construct all possible code words by multiplying our generator matrix by all combinations of information bits. For two information bits, the possible combinations are given by $\{00, 01, 10, 11\}$. The resulting four different code words $\underline{c}_m$ are given by

$$00 \rightarrow \underline{c}_1 = (0 \quad 0) \begin{pmatrix} 1 & 0 & | & 1 & 0 & 0 \\ 0 & 1 & | & 1 & 1 & 1 \end{pmatrix}$$

$$= \{0 \cdot 1 + 0 \cdot 0, \, 0 \cdot 0 + 0 \cdot 1, \, 0 \cdot 1 + 0 \cdot 1, \, 0 \cdot 0 + 0 \cdot 1, \, 0 \cdot 0 + 0 \cdot 1\}$$

$$= \{0\,0\,0\,0\,0\}$$

$$01 \rightarrow \underline{c}_2 = (0 \quad 1) \begin{pmatrix} 1 & 0 & 1 & 0 & 0 \\ 0 & 1 & 1 & 1 & 1 \end{pmatrix}$$

$$= \{0 \cdot 1 + 1 \cdot 0, \, 0 \cdot 0 + 1 \cdot 1, \, 0 \cdot 1 + 1 \cdot 1, \, 0 \cdot 0 + 1 \cdot 1, \, 0 \cdot 0 + 1 \cdot 1\}$$

$$= \{0\,1\,1\,1\,1\}$$

$$10 \rightarrow \underline{c}_3 = (1 \quad 0) \begin{pmatrix} 1 & 0 & 1 & 0 & 0 \\ 0 & 1 & 1 & 1 & 1 \end{pmatrix}$$

$$= \{1 \cdot 1 + 0 \cdot 0, \, 1 \cdot 0 + 0 \cdot 1, \, 1 \cdot 1 + 0 \cdot 1, \, 1 \cdot 0 + 0 \cdot 1, \, 1 \cdot 0 + 0 \cdot 1\}$$

$$= \{1\,0\,1\,0\,0\}$$

$$11 \rightarrow \underline{c}_4 = (1 \quad 1) \begin{pmatrix} 1 & 0 & 1 & 0 & 0 \\ 0 & 1 & 1 & 1 & 1 \end{pmatrix}$$

$$= \{1 \cdot 1 + 1 \cdot 0, \, 1 \cdot 0 + 1 \cdot 1, \, 1 \cdot 1 + 1 \cdot 1, \, 1 \cdot 0 + 1 \cdot 1, \, 1 \cdot 0 + 1 \cdot 1\}$$

$$= \{1\,1\,0\,1\,1\}, \tag{8.9}$$

where we remember the binary arithmetic

$$0 + 0 = 0 \tag{8.10}$$
$$0 + 1 = 1 \tag{8.11}$$
$$1 + 0 = 1 \tag{8.12}$$
$$1 + 1 = 0 \quad \leftarrow \text{ the funny one} \tag{8.13}$$

and

$$0 \cdot 0 = 0 \tag{8.14}$$
$$0 \cdot 1 = 0 \tag{8.15}$$
$$1 \cdot 0 = 0 \tag{8.16}$$
$$1 \cdot 1 = 1. \tag{8.17}$$

Because our code words are linear, the sum of any two code words is a code word that is in the set. The sum of a code word and the all-zero code word is the original code word. As another example, we see that $\underline{c}_3 + \underline{c}_4$ is given by $\underline{c}_2$:

$$\underline{c}_3 + \underline{c}_4 = \{1\,0\,1\,0\,0\} + \{1\,1\,0\,1\,1\}$$
$$= \{0\,1\,1\,1\,1\} = \underline{c}_2. \tag{8.18}$$

The sum of codewords $\underline{c}_m$ and $\underline{c}_l$ associated with information bits $\underline{x}_m$ and $\underline{x}_l$ is given by

$$\underline{c}_m = \underline{x}_m \, \mathbf{G}$$
$$\underline{c}_l = \underline{x}_l \, \mathbf{G}$$
$$\underline{c}_m + \underline{c}_l = \underline{x}_m \, \mathbf{G} + \underline{x}_l \, \mathbf{G}$$
$$= (\underline{x}_m + \underline{x}_l) \, \mathbf{G}$$
$$= \underline{x}_j \, \mathbf{G}, \tag{8.19}$$

where the sum information bit sequence is given by $\underline{x}_j = \underline{x}_m + \underline{x}_l$. Because all information bit sequences are allowed, the sum of information sequence vectors is another information sequence vector. Thus, the sum of code words is another code word. Another way to state this is that code words form a closed set under addition. Also, we observe from the form of the generator matrix that its rows are code words. All code words are constructed from a linear combination of these code words expressed in the generator matrix.

One of the important characteristics of an error correcting code is the distance between the code words. Bit error rate (BER) performance is typically driven by the smallest distance. We measure this distance in terms of the Hamming distance, which is the minimum number of bit changes to change one sequence to another. For our toy code, the Hamming distance between all code words is given by

$$d(\underline{c}_1, \underline{c}_2) = 4$$
$$d(\underline{c}_1, \underline{c}_3) = 2 \Leftarrow \text{shortest distance}$$
$$d(\underline{c}_1, \underline{c}_4) = 4$$
$$d(\underline{c}_2, \underline{c}_3) = 4$$
$$d(\underline{c}_2, \underline{c}_4) = 2 \Leftarrow \text{shortest distance}$$
$$d(\underline{c}_3, \underline{c}_4) = 4. \tag{8.20}$$

In this case, the smallest distance is somewhat annoying because not all single bit errors can be unambiguously corrected. Consider a bit error that moves $\underline{c}_1$, which we denoted $\underline{\tilde{c}}_1$, closer to $\underline{c}_3$. It would be unclear if the best error correcting choice is $\underline{c}_1$ or $\underline{c}_3$ because they are both Hamming distance 1 from $\underline{\tilde{c}}_1$. If the minimum distance for the code had been 3, then any single bit error could be corrected. We see an example of this effect in Figure 8.3. Here, there is a true code word and two other code words with Hamming distances of 2 and 3 from the true code word. If the single bit error moves toward the code word that is distance 3 from the true code word, the error can be mitigated. However, if the the bit error moves toward the code word that is Hamming distance 2 from the true code word, it is unclear which code word is the better choice for correction.

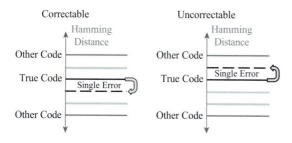

Figure 8.3 Notional example of single bit error and Hamming distances between code words.

8.4   Parity Check Matrix

We can construct a matrix that identifies a small number of errors given some observation of a number of observed bits. To do this, we assume that we have made hard decisions. We expect better performance if we pass likelihoods rather than hard decisions onto our decoder, but this approach is beyond our discussion.

If we are using a linear error correction code, we can perform this small number of error checks by using the parity check matrix $\mathbf{H}$. If the product of the observed bit vector $\underline{y}$ and the parity check matrix is the zero vector, then there are no detectable errors:

$$\underline{y}\,\mathbf{H}^T = \underline{0}. \tag{8.21}$$

The transpose is because of convention. We find this matrix by observing that the product of any valid code word and the matrix gives the zero vector, so that the product is given by

$$\underline{c}_m\,\mathbf{H}^T = \underline{0}, \quad \mathbf{H} \in \mathbb{Z}_2^{(n-k)\times n} \tag{8.22}$$
$$\underline{x}_m\mathbf{G}\,\mathbf{H}^T = \underline{0}. \tag{8.23}$$

We note that the relationship is always satisfied if $\mathbf{G}\,\mathbf{H}^T = \mathbf{0}$, where $\mathbf{0}$ is a matrix of zeros. For the systematic codes, this can be satisfied if

$$\mathbf{H} = (-\mathbf{P}^T \quad \mathbf{I}) \tag{8.24}$$

because

$$\mathbf{G}\,\mathbf{H}^T = (\mathbf{I} \quad \mathbf{P})(-\mathbf{P}^T \quad \mathbf{I})^T$$
$$= -\mathbf{P} + \mathbf{P} = \mathbf{0}. \tag{8.25}$$

If we are using binary or, equivalently, GF(2) symbols, then the parity check matrix is given by

$$\mathbf{H} = (\mathbf{P}^T \quad \mathbf{I}); \text{ for binary} \tag{8.26}$$

because $-\mathbf{P} = \mathbf{P}$ for GF(2).

8.5   Graph Representation of Parity Check Matrix

The parity check matrix can be represented as a bipartite graph that relates parity check operations vertices to individual symbol nodes of code words. For our toy code that was introduced in Equation (8.8), the parity check matrix is given by

$$\mathbf{H} = (\mathbf{P}^T \quad \mathbf{I}); \text{ for binary}$$
$$= \begin{pmatrix} 1 & 1 & 1 & 0 & 0 \\ 0 & 1 & 0 & 1 & 0 \\ 0 & 1 & 0 & 0 & 1 \end{pmatrix}. \tag{8.27}$$

The corresponding graph representation is depicted in Figure 8.4. This graph representation is sometimes useful for discussing performance and approaches for decoding. In particular, for soft decoders, which we will not discuss in detail, likelihoods are evaluated and modified at either set of the vertices associated with the code word symbols or parity check vertices. The iterative soft decoder passes these likelihoods between the two sets of vertices until likelihoods converge. The result of this iteration is an estimate of the original code word.

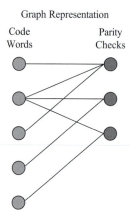

Graph Representation

Code   Parity
Words   Checks

Figure 8.4 Bipartite graphical representation of the parity check matrix.

8.6   Syndrome

The parity check matrix is designed to produce a vector of zeros if it is multiplied by a valid code word. The resulting vector from the product of the observed bits and the parity check matrix is the syndrome. If there are symbol errors, the syndrome can be used to attempt to correct the errors.

We denote a sequence of observed binary symbols $\underline{y} \in \mathbb{Z}_2^{1 \times n}$, for which we made hard decisions,

$$\underline{y} = \underline{c}_m + \underline{e},$$
(8.28)

where $\underline{c}_m$ is the transmitted code word and $\underline{e}$ indicates a vector of bit errors.

The syndrome is then indicated by

$$\underline{s} = \underline{y}\,\mathbf{H}^T \in \mathbb{Z}_2^{1 \times (n-k)}.$$
(8.29)

By explicitly introducing the errors, we observe

$$\underline{s} = (\underline{c}_m + \underline{e})\,\mathbf{H}^T$$
$$= \underline{c}_m\,\mathbf{H}^T + \underline{e}\,\mathbf{H}^T$$
$$= \underline{0} + \underline{e}\,\mathbf{H}^T = \underline{e}\,\mathbf{H}^T$$
(8.30)

so that, for this linear code, the syndrome is independent of the code word and is only a function of the error vector $\underline{e}$. Consequently, under the correct conditions, we can draw a correspondence between particular syndrome vectors and their corresponding error vectors. Unfortunately, it is not immediately clear which errors correspond to a given syndrome. Finally, we can hopefully correct the errors and recover the original code word. From this corrected code word, we extract the original information bits.

Not all symbol errors can be corrected, so our first problem is to determine which errors we can correct. We can construct a table of correctable errors, that is, a syndrome table. For our code, there are four possible code words; thus, we have four columns. We have three parity bits; thus, we have $2^3 = 8$ possible syndrome sequences and corresponding potentially correctable error sequences. For the block code suggested in Reference [2], we hypothesize various likely error sequences and calculate the corresponding syndromes:

$\underline{c}_1$ or $\underline{e}$	$\underline{c}_2$	$\underline{c}_3$	$\underline{c}_4$	
00000	01111	10100	11011	syndrome $= 000$
10000	11111	00100	01011	syndrome $= 100$
01000	00111	11100	10011	syndrome $= 111$
00010	01101	10110	11001	syndrome $= 010$
00001	01110	10101	11010	syndrome $= 001$
11000	10111	01100	00011	syndrome $= 011$
10010	11101	00110	01001	syndrome $= 110$
10001	11110	00101	01010	syndrome $= 101$

$$(8.31)$$

Because we have an all-zero code word $\underline{c}_1$, any difference from that code word is also the error vector. Note that because we have a poor code, not all single errors are correctable. In particular, an error in the third symbol for $\underline{c}_1$ produces the same sequence as an error for the first symbol for $\underline{c}_3$. By luck of the way we constructed the table, we correct the $\underline{c}_3$ error and give up on $\underline{c}_1$. Given this weak code, we cannot correct one possible single bit error and most double errors.

To recover our best estimate of the original code word $\hat{\underline{c}}$, we evaluate the syndrome to identify the corresponding error sequence from the first column of syndrome table $\underline{e}$ and correct the observed symbols with this information. For binary symbols, we simply add the error vector to the observation vector:

$$\hat{\underline{c}} = \underline{e} + \underline{y}. \qquad (8.32)$$

For our systematic code, we can read off the first two symbols to recover the information bits.

8.7   Hamming Codes

While modern approaches, such as turbo codes and low-density-parity-check (LDPC) codes, are much more capable, we focus on linear block codes. Hamming codes are a useful set of simple block codes [15]. Hamming codes indicate a class of systematic linear block codes that can correct any single bit error in an observed code word. They have minimum Hamming distance of $d_{min} = 3$, a length $n = 2^m - 1$, and a number of information bits given by $k = 2^m - m - 1$, for $m > 1$. An example Hamming code is the $(n, k) = (7, 4)$ code with coding rate $4 / 7$, which is constructed with the generator matrix $\mathbf{G}$,

$$\mathbf{G} = \left(\begin{array}{cccc|ccc} 1 & 0 & 0 & 0 & 1 & 1 & 0 \\ 0 & 1 & 0 & 0 & 0 & 1 & 1 \\ 0 & 0 & 1 & 0 & 1 & 0 & 1 \\ 0 & 0 & 0 & 1 & 1 & 1 & 1 \end{array} \right), \qquad (8.33)$$

with corresponding parity check matrix $\mathbf{H}$:

$$\mathbf{H} = \left(\begin{array}{cccc|ccc} 1 & 0 & 1 & 1 & 1 & 0 & 0 \\ 1 & 1 & 0 & 1 & 0 & 1 & 0 \\ 0 & 1 & 1 & 1 & 0 & 0 & 1 \end{array} \right). \qquad (8.34)$$

8.8   Convolutional Codes

While the block code segments information bits into blocks and applies this operation in groups of this size, the convolutional code performs encoding in a streaming fashion so that the same code can be applied to arrays of information bits of different sizes. Convolutional codes are widely used but are not typically particularly strong, where strong means getting close to the Shannon capacity. Interestingly, by using two convolutional codes and an interleaver, one can create a turbo code that is very strong.

The fundamental tool of the convolutional code is the shift register. As a new bit or bits are shifted into the register, combinations of bits are constructed by addition if we are using GF(2) – or, equivalently, by using *exclusive-or*s (XORs) – to provide the encoded bits. If the output of the XORs is applied in some way to the input of the shift register, then we call this a *recursive convolutional code*; otherwise it is a *non-recursive convolutional code*. If an information bit is also passed directly to the output sequence, then we term it a *systematic code*, similar to the block codes that we discussed previously. It is convenient to represent these codes mathematically, pictorially, and graphically.

A mathematical way of describing these codes is by recognizing that the effects of the shift register and XORs ($\oplus$) can be expressed as a convolution in GF(2) or, equivalently, in binary. Thus, we have the codes' name. Because our code must convert some number of information bits to a larger number of output bits, the code has multiple convolutional paths. Each of these paths is produced by a different function of current and past inputs. The jth output path has its own transfer function h_k^j for its kth element of memory. Thus, we have

$$y_m^j = \sum_k h_k^j x_{m-k}, \qquad (8.35)$$

where y_m^j is the output of the mth bit convolutional encoder's jth path. We can consider this relationship between the input and output bits by using the Z-transform of the convolution while reminding ourselves that the calculations are binary. The transform of the jth path is then give by H^j:

$$H^j = \sum_k P_k^j z^{-k} \qquad (8.36)$$

for a non-recursive convolutional code, and

$$H^j = \frac{\sum_k P_k^j z^{-k}}{\sum_n Q_n^j z^{-n}} \qquad (8.37)$$

for a recursive convolutional code, where the z^{-1} indicates a unit delay. We can implement the generation of these codes by using shift registers and XORs. The XOR implements binary [GF(2)] addition. We depict examples of recursive and non-recursive codes in Figure 8.5. For these examples, the codes are rate 1/2, that is, 2 bits are produced for every information bit, and the constraint length is four, which we indicate by $K = 4$, which requires three memory registers. In this case, the recursive code is systematic, which means the information bits are explicitly represented in the code output.

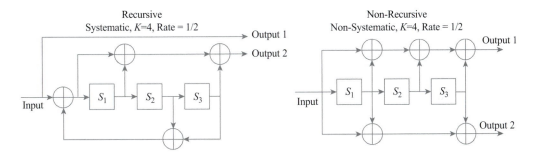

Figure 8.5 Examples of simple recursive and non-recursive convolutional shift register coding implementations.

It is pedagogically useful to use an extremely simple code to understand the workings of encoding and decoding of convolutional codes. Our example uses a rate-1/2, non-recursive, constraint length $K = 3$, convolutional code. We depict the shift register encoding representation of our example code in Figure 8.6.

Because the shift register has finite memory, there are a finite number of possible states, that is, a finite number of sequences of shift register values. For our simple example, we can easily enumerate all the states (Table 8.1). The previous information bit is stored in S_1. The bit prior to that one is stored in S_2. All possible output sequences are determined by the state of S_1, S_2 (for which we use the notation S_1 / S_2) and the input bit. As the information bits flow through the encoder, each input information bit is turned into two output bits for our rate-1/2 code. The output

Table 8.1 Overview of states for simple example convolutional code depicted in Figure 8.6.

New Bit	S_1/S_2	C_1	C_2	S_1'/S_2'
0	0 / 0	0	0	0 / 0
1		1	1	1 / 0
0	0 / 1	1	0	0 / 0
1		0	1	1 / 0
0	1 / 0	1	1	0 / 1
1		0	0	1 / 1
0	1 / 1	0	1	0 / 1
1		1	0	1 / 1

Non-Recursive, K=3, Rate = 1/2

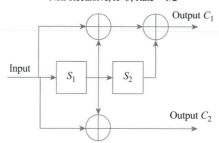

Figure 8.6 Example using a rate-1/2, non-recursive, constraint length $K = 3$, convolutional code.

bits are denoted C_1 and C_2. At the next step in time, the new states are given by $S_1' = $ new bit b and $S_2' = S_1$.

We can represent this information on a trellis diagram that represents all possible state transitions: $S_1, S_2 \rightarrow S_1', S_2'$. The state transition is caused by the new bit

Trellis

s_1 / s_2 $b \rightarrow (C_1, C_2)$

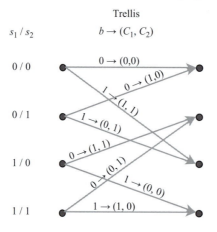

Figure 8.7 Trellis diagram for our example that uses a rate-1/2, non-recursive, constraint length $K = 3$, convolutional code.

that is input into the shift register. We depict this transition in Figure 8.7. We can define the possible output bits C_1, C_2 based upon the initial state S_1 / S_2 and the new information bit b.

We consider the encoding of the information bit sequence 11001, where we start from the left and move right. We start with the states 0 / 0. By moving through the trellis depicted in Figure 8.8, the input sequence is transformed to the output sequence 11001 $\rightarrow$ 11 00 01 10 11.

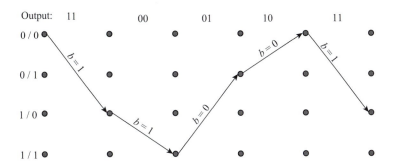

Figure 8.8 Rate-1/2, non-recursive, constraint length $K = 3$, convolutional code encoding example with input sequence sequence 11001.

8.8.1   Viterbi Decoder

The Viterbi decoder [16] finds the maximum likelihood path through a trellis. It works for both hard decisions (by accumulating Hamming distances) and soft decisions (by accumulating symbol likelihoods). The Viterbi decoder performs the following operations:

- Start in the all-zero state.
- Evaluate the distance between the predicted observed symbols and the observed symbols for each possible state-to-state transition:
  - Hamming distance for hard decisions;
  - typically log-likelihood for soft decisions.
- Accumulate the sum of the distances up to the current state transition.
- Keep the input path with the lowest accumulated distance by considering the possible inputs to a given state.
- At the last state transition, select the path with the smallest accumulated distance.
- Trace the path back to the start to find the best state transitions and, consequently, the information bits.

When there are multiple paths with the same accumulated distance, it is an indication that we need better input SNR or a stronger code, and we have no choice but to just pick one path.

We consider the example with information bits 11001 that produce the encoded sequence 11 00 01 10 11 detailed in Figures 8.6 and 8.8. For this example, we assume that the receiver employs hard decisions. We flip the fifth encoded bit so that the hard decision of the received sequence is given by 11 00 01 10 11. For our example, we start in the all-zero state 0 / 0, which is typical. From Figure 8.7 and if the information bit was a 0 or 1, possible transitions are to 0 / 0 and 1 / 0, respectively. Given that we observed the two symbols 11, we can calculate the Hamming distance between the observed sequence and the possible sequences coming from the 0 / 0. Again from Figure 8.7, we know that the two possible encoded sequences are 00 and 11, which then transition to states that correspond to Hamming distances of 2 and 0, respectively. We now have two possible starting states. From each of these states, we can transition to all of the four possible states. We can calculate the Hamming distances d for each of these distances given that we observed 00, as we depict in Figure 8.9. We can also accumulate the sum distances s over possible paths. During the next state transition, we get to the point where all transitions allowed by the trellis (as seen in Figure 8.7) can be followed. In this case, we observe 11 (which should be 01, but there was a bit error). We calculate all eight possible Hamming distances, and we accumulate the sum distances. We now enter a new phase of the calculation. There are multiple paths entering each state in the trellis diagram. In our example, two paths enter each state. There is no value in keeping a path with a higher accumulated distance s, so we can trim that path. As an example, the sum distances entering the 0 / 0 state are $s = 4$ and $s = 3$. There is no reason to keep the $s = 4$ path, so we trim it. Next, we repeat the process of measuring the distance, calculating the sum, and trimming the paths of higher accumulated distance until we get to the end of the observed bits. At this point, we see that the path transitioning from the 0 / 0 to 1 / 0 state has the lowest accumulated distance. We now follow this path back to the first 0 / 0 to find the best path. From Figure 8.7, we see that the best information bit

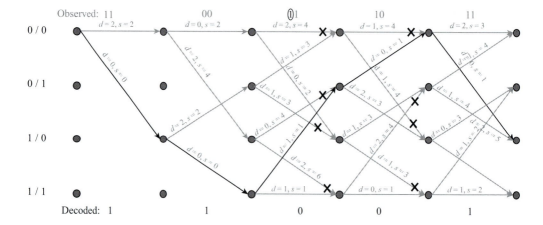

Figure 8.9 Rate-1/2, non-recursive, constraint length $K = 3$, convolutional code encoding example with input sequence 11001. Hamming distances encoded symbols for possible transitions and the corresponding observed symbols are represented by d. The accumulated sums of distances are represented by s.

sequence associated with the state transitions of this path is given by 11001, which is correct. The Viterbi decoding successfully corrected the symbol error.

There are a remarkable number of extensions to the basic Viterbi algorithm that we discussed in References [17, 18]. However, these discussions are beyond the scope of our text.

Problems

8.1 By using simulation, evaluate the BER for QPSK symbols in Gaussian noise as a function of the signal-to-noise ratio (SNR). Compare this BER to a simple repetition code for which the symbols are repeated three times and the energy of each transmitted symbol is divided by 3. For the repetition code, vote across the bits to decode.

8.2 After encoding, evaluate the total number of bits and the coding rate for the following block codes of information bits and parity bits.

n_{info}	n_{parity}
1	2
4	3
11	4
26	5

8.3 What is the spectral efficiency of
(a) rate 3/4 code with BPSK modulation?
(b) rate 1/2 code and QPSK modulation?
(c) rate 1/3 code with 8-PSK modulation?
(d) rate 4/7 code with 16-QAM modulation?

8.4 What is the Hamming distance between the sequence $(1, 0, 0, 1)$ and
(a) $(1, 0, 0, 0)$?
(b) $(0, 0, 0, 0)$?
(c) $(0, 0, 1, 1)$?
(d) $(1, 1, 0, 0)$?
(e) $(1, 1, 1, 1)$?

8.5 By using the (7,4) Hamming code generator matrix presented in this chapter, construct the encoded bits for
(a) $(1, 0, 0, 0)$
(b) $(0, 0, 0, 0)$
(c) $(0, 0, 1, 1)$
(d) $(1, 1, 0, 0)$
(e) $(1, 1, 1, 1)$

8.6 By using the (7,4) Hamming code generator matrix, construct the parity check matrix. Show how you made it. What is the coding rate of this code?

8.7 Show that any Hamming code encoded sequence of information bits inputs has zero syndrome. Note that this is true of Hamming codes of any size, not just the one that we discussed.

8.8 Construct a syndrome table for an implementation of the (7,4) Hamming code. You might want to do this in software, but you can do it by hand.

8.9 In simulation, construct an implementation and simulation of a QPSK modulation of the (7,4) Hamming code. Evaluate the BER as a function of SNR for a flat-fading channel with a known channel coefficient.

8.10 Consider a (6,2) binary, linear, systematic, block code.
(a) If $(1, 1, 0, 0, 1, 1)$ and $(1, 0, 1, 1, 0, 0)$ are code words, construct the complete set of code words.
(b) Construct the generator matrix for the code.
(c) Construct the parity check matrix.
(d) Evaluate the minimum Hamming distance for the code.
(e) Can this code correct all single bit errors? Why or why not?

8.11 In simulation, develop and implement the encoding, BPSK modulation, hard-decision demodulation, and Viterbi decoding for the example rate 1/2, $k=3$, convolutional code discussed in this chapter. Evaluate the BER performance by transmitting pseudorandom information bit sequences of length 100. Sweep across symbol SNR of 0 dB to 10 dB. Average over 100 throws of the 100 information bits at each SNR. Plot the BER as a function of SNR.

8.12 In simulation, develop and implement the encoding, BPSK modulation, soft-decision demodulation, and Viterbi decoding for the example rate 1/2, $k=3$, convolutional code discussed in this chapter. Use the sum squared distance from each possible BPSK symbol (which is proportional to the log-likelihood) to determine distance. Evaluate the BER performance in flat-fading additive Gaussian noise by transmitting pseudorandom information bit sequences of length 100. Sweep across symbol SNR of 0 dB to 10 dB. Average over 100 throws of the 100 information bits at each SNR. Plot the BER as a function of SNR.

9 Acquisition and Synchronization

In this chapter, we discuss the ideas of signal acquisition and radio-to-radio synchronization in both time and frequency. We address the critical question: Is anyone out there? We discuss the uncertainty in time and frequency alignment between radios. We introduce and analyze the performance of multiple signal acquisition techniques: energy, cross-correlation, normalized inner product, and autocovariance detectors. We develop the maximum likelihood estimators for temporal and spectral synchronization for single-carrier approaches. Finally, we also introduce a temporal synchronization approach for an OFDM symbol.

While there is some semantical ambiguity in discussing acquisition, fundamentally the term indicates that the existence of the communications signal of interest has been detected. For most communications systems, detection must be followed by the additional task of frequency and fine-time synchronization. The synchronization process enables the transmitter and receiver to agree on carrier frequency and packet timing. For nearly all systems, phase synchronization (if required) is performed by the receiver. This is often termed *channel compensation*, which was discussed in Sections 6.3.1 and 7.3. In high-signal-to-noise ratio (SNR) regimes, for some modulation schemes, synchronization can be performed without explicit knowledge of the symbol sequences. However, many approaches explicitly employ training or pilot data to aid these processes.

9.1 Time and Frequency Uncertainty

Acquisition is an extremely important component of communications systems. We can imagine that a radio is turned on. It might have an approximate sense of time, but typically it is not sufficiently accurate to know when some data is expected. Furthermore, even if the timing were known, there may be no transmitter within a reasonable distance. The receiver must first determine if a transmitter has sent a signal and the timing and center frequency of the signal. We call this process *acquisition*.

Implicit in essentially all discussions up to this point is the assumption that the receiver knows at what frequency and time to listen. While communications could be prearranged, it is difficult to maintain precise frequency and timing alignment. Commercial-grade local oscillators have frequency accuracy on the order of 10^{-6}. For each 10^6 cycles, there is typically one too many or too few cycles. If we were transmitting at a carrier frequency of 1 GHz, the transmitter and receiver would typically disagree on the carrier frequency $\Delta f \sim 10^{-6} \cdot 1$ GHz, that is, by a couple of kilohertz. Physical Doppler-frequency offsets can have a similar effect. For non-relativistic speeds, the Doppler-frequency offset is given by

$$\Delta f = \frac{v}{c_{\text{light}}} f_c, \tag{9.1}$$

where v is the velocity, c_{light} is the speed, and f_c is the carrier frequency. For a 1 GHz carrier frequency and a 30 m/s vehicle speed, the Doppler shift produces a frequency offset of about 100 Hz.

If uncorrected, the frequency offset would limit the period of time over which we can maintain coherence. If a transmitted constellation was aligned at time zero ($t = 0$) when observed at the receiver, the constellation would rotate by

$$\theta = \arg e^{i 2\pi \Delta f t}$$
$$= 2\pi \Delta f t, \tag{9.2}$$

where Δf is the difference in the carrier frequency assumed by the transmitter and receiver. The sensitivity to this potential angle error is dependent upon how the processing is performed and on the operating spectral efficiency. As an example, consider the use of quadrature-phase-shift keying (QPSK) with a $\Delta f = 1$ kHz. To prevent the phase error from significantly contributing to decoding errors, it needs to be small compared to half of the angular separation of the constellation points. For example, the phase error needs to be small compared to $\pi/4$ for QPSK. Consequently, the processing without calculating a new phase correction would be given by

$$2\pi \Delta f t \ll \frac{\pi}{4}$$
$$t \ll \frac{1}{8 \Delta f} = 125 \, \mu s. \tag{9.3}$$

Depending upon the bandwidth of the signal and the frequency offset, acceptable duration might not correspond to that many symbols. Furthermore, higher-order constellations will be even more sensitive. This effect motivates the synchronization of the frequencies between the transmitter and receiver. Compensating for the frequency offset does not need to be performed in hardware. In modern systems, the frequency offset is usually digitally implemented at complex baseband.

Timing offsets can be even more problematic. The receiver may be completely unaware of the impetus for a transmitter to send a message. This impetus might be the result of some external stimuli – for example, you want to make a call. How could the receiver possibly know when this will happen? Even if the time was carefully prearranged, the error in the local oscillator would cause the timing to drift. Consider an agreement to transmit and receive one day (86,400 s) later. By assuming a relative local oscillator error of 10^{-6}, the receiver would be misaligned by 86 ms, which may not seem like much for human perception, but for a communications system this offset corresponds to many symbols. For a symbol rate of 1 Msym/s, the signal would be misaligned by 86,400 symbols. For symbols to make sense, the misalignment in time must be small compared to a symbol duration.

9.2 Acquisition

The initial step in communicating is determining whether there is a transmitter trying to send a signal to you. There are numerous techniques to perform this acquisition. They all exploit some difference between a model for background noise and a model for a signal plus noise. Most modern systems exploit specific knowledge of the transmitted waveform.

9.2.1 Energy Detector

If the signal is received with relatively high SNR, the receiver can detect that a signal exists by observing a change in receiver power. For a row vector of N observed complex baseband samples $\underline{z} \in \mathbb{C}^{1 \times n_s}$, the energy of the sequence is given by

$$E = \|\underline{z}\|^2 = \sum_{m=1}^{n_s} \|\underline{z}_m\|^2$$

$$= \sum_{m=1}^{n_s} (\Re\{\underline{z}_m\}^2 + \Im\{\underline{z}_m\}^2). \tag{9.4}$$

We consider two hypotheses: first, there is just thermal noise present, $\mathcal{H}_0$, of known variance σ_n^2. To be clear, the noise variance is not always known, but we will ignore that subtlety in this discussion. Second, there is thermal noise plus a Gaussian signal, $\mathcal{H}_1$. The probability density for the thermal noise hypothesis $\mathcal{H}_0$ is given by

$$p(\underline{z}|\mathcal{H}_0) = \frac{1}{\pi^N \sigma^{2n_s}} e^{-\|\underline{z}\|^2/\sigma_n^2}. \tag{9.5}$$

The probability density function for E is given by a complex χ^2 distribution, described in Section 14.1.10. This distribution is given by

$$p(E|\mathcal{H}_0) = p_{\chi^2}^{\mathbb{C}}(E; N, \sigma_n^2)$$

$$= \frac{E^{n_s-1}}{(\sigma_n^2)^{n_s} \Gamma(n_s)} e^{-\frac{E}{\sigma_n^2}}, \tag{9.6}$$

where $\Gamma(n)$ is the gamma function described in Section 13.10.2. For the signal plus noise hypothesis with a complex Gaussian signal with variance σ_s^2, the probability density function for E is given by

$$p(E|\mathcal{H}_1) = p_{\chi^2}^{\mathbb{C}}(E; n_s, [\sigma_n^2 + \sigma_s^2])$$

$$= \frac{E^{Nn_s-1}}{(\sigma_n^2 + \sigma_s^2)^{n_s} \Gamma(n_s)} e^{-\frac{E}{\sigma_n^2+\sigma_s^2}}. \tag{9.7}$$

We set a threshold η on E such that we declare detection if $E \geq \eta$. On occasion, the noise will fluctuate such that a false detection occurs. This is a false alarm, which is sometimes called a *false-positive* or *type-one error*. Conversely, sometimes the observed energy will fluctuate down because of the sum of the noise with the signal so that the detection will be missed. This error is sometimes called a *false-negative* or *type-two error*. We characterize the performance by comparing the probability of detection P_D under $\mathcal{H}_1$ versus the probability of false alarm P_{FA} under $\mathcal{H}_0$.

The probability of detection P_D can be evaluated with

$$P_D = \int_\eta^\infty dE \, p(E|\mathcal{H}_1)$$

$$= 1 - \text{CDF}(\eta|\mathcal{H}_1), \tag{9.8}$$

which is given by the complementary cumulative distribution function for this hypothesis. The cumulative distribution function (CDF) for the complex χ^2 distribution, discussed in Section 14.1.10, is given by

$$\text{CDF}(\eta|\mathcal{H}_1) = P_{\chi^2}^C(\eta; n_s, [\sigma_n^2 + \sigma_s^2])$$

$$P_D = 1 - \frac{1}{\Gamma(n_s)} \gamma\left(n_s, \frac{\eta}{\sigma_n^2 + \sigma_s^2}\right), \tag{9.9}$$

where $\gamma(n, x)$ is the lower incomplete gamma function described in Section 13.10.2.

Similarly, the probability of a false alarm P_{FA} is given when the observed energy of just the noise fluctuates above the threshold η. The probability is given by

$$P_{FA} = \int_\eta^\infty dE \, p(E|\mathcal{H}_0)$$

$$= 1 - \text{CDF}(\eta|\mathcal{H}_0)$$

$$= 1 - \frac{1}{\Gamma(n_s)} \gamma\left(n_s, \frac{\eta}{\sigma_n^2}\right). \tag{9.10}$$

In Figure 9.1, we display the receiver operating characteristic (ROC) performance of an energy detector for a signal with per-sample SNR of 0 dB. We observe the probability of detection versus probability of false-alarm performance for a single sample and for 10 samples, the latter of which performs dramatically better.

9.2.2   Cross-Correlation Detector

Probably the most common approach to performing acquisition is to measure correlation with a known transmitted reference, training, or pilot sequence. For some known sequence $\underline{s} \in \mathbb{C}^{1 \times n_s}$, we ask how similar it is to the current observed data $\underline{z}$. We apply this test repeatedly as data flows through our test observed data $\underline{z}$. The most common approach is to employ a correlator. We have a two-hypothesis test: first, the signal is not there, H_0; second, the signal is there and properly aligned in time, H_1. In the third scenario, the signal is approximately there but misaligned. We ignore this third scenario for the moment.

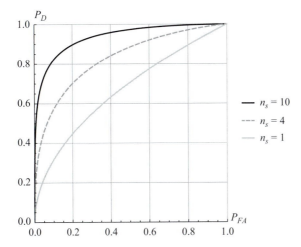

Figure 9.1 Receiver operating characteristic performance of an energy detector for SNR of 0 dB. Single-, 4-, and 10-sample performance is compared.

Our model for the received signal under the first hypothesis H_0 assumes that there is only noise given by

$$\underline{z} = \underline{n},$$ (9.11)

where the entries in $\underline{n} \in \mathbb{C}^{1 \times n_s}$ are independently drawn from a zero-mean circularly symmetric complex Gaussian distribution (see Section 14.1.6). For the sake of simplifying the discussion, we assume independent unit-variance noise, $\sigma^2 = 1$, so the covariance is given by the identity matrix $\mathbf{R} = \mathbf{I}$. The model for the second hypothesis is given by

$$\underline{z} = a\underline{s} + \underline{n},$$ (9.12)

where a is some constant that involves the amplitude product of the amplifier gain and channel loss. We assume the reference is normalized so that $\|\underline{s}\|^2 = n_s$. Under these definitions, the SNR is given by a^2. A natural test is then to evaluate the magnitude squared of the inner product of the test data and the reference sequence. We use the observed data and our knowledge of the transmitted signal to construct a function. If the value of this function is greater than some threshold, then we declare a detection. This function is sometimes called a *detection test statistic*, which is a slightly confusing nomenclature. The correlation approach gives us a detection test statistic ϕ such that ϕ is given by

$$\phi = \|\underline{z}\,\underline{s}^H\|^2$$
$$= \|a\underline{s}\,\underline{s}^H + \underline{n}\,\underline{s}^H\|^2$$
$$= \|a\,n_s + \check{n}\|^2,$$ (9.13)

where the sum of the n_s complex noise samples is a new random scalar $\check{n} = \underline{n}\,\underline{s}^H \in \mathbb{C}$. The variance of this new random variable is $\mathrm{Var}\{\check{n}\} = n_s$. We divide the detection test statistic by n_s to produce a new detection test statistic that the variance of noise component is $\text{Var}\{\tilde{n}\} = 1$, given by $\tilde{n} = \check{n}/\sqrt{n_s}$. The new detection test statistic is given by

$$\tilde{\phi} = \frac{1}{n_s}\|a\,n_s + \mathbf{n}\,\underline{s}^H\|^2$$
$$= \|a\,\sqrt{n_s} + \tilde{n}\|^2 . \tag{9.14}$$

The rescaling is a monotonic transformation of the previous version of the detection test statistic. This transformation does not affect the shape of the ROC performance, but it does change the specific threshold that provides the given probabilities of detection and of false alarm. Consequently, the performance characteristic of the detection test statistic remains the same. We could use this form so the distributions for the two hypotheses would be χ^2 and noncentral χ^2 distributions. However, to exercise some other distributions, we apply another transform to the detection test statistic again by taking the square root of the previous version:

$$y = \sqrt{\tilde{\phi}}$$
$$= \|a\,\sqrt{n_s} + \tilde{n}\| , \tag{9.15}$$

which is once again a monotonic transformation.

We consider the probability density function of the detection test statistic y under the signal-aligned and null hypotheses. The distribution for $\tilde{\phi}$ under the null (just noise) hypothesis is the central χ^2 distribution with two real (or one complex) degrees of freedom, as discussed in Section 14.1.10. If we use the form for y, the probability density function is given by a Rayleigh distribution, as we define in Section 14.1.7. This distribution is given by

$$p_{Ray}(y)\,dy = \begin{cases} 2\,y\,e^{-y^2}\,dy & ; y \geq 0 \\ 0\,dy & ; \text{ otherwise.} \end{cases} \tag{9.16}$$

The probability of a false alarm is given by the probability that, under the null hypothesis, the detection test statistic y exceeds the threshold $\eta \geq 0$:

$$P_{FA} = \int_{\eta}^{\infty} dy\, p_{Ray}(y) = e^{-\eta^2} . \tag{9.17}$$

Similarly, to determine the probability of detection, we consider the probability that the detection test statistic y exceeds the threshold η. The distribution for $\tilde{\phi}$ under the aligned-signal hypothesis is given by a noncentral χ^2 distribution with two degrees of freedom. The distribution for y is given by a Rician distribution, which we discuss in Section 14.1.11, and is given by

$$p_{Rice}(y)\,dy = \begin{cases} 2y\,I_0\left(\frac{2\,\|a\,\sqrt{n_s}\|\,y}{\sigma^2}\right)e^{-(y^2 + \|a\,\sqrt{n_s}\|^2)/\sigma^2}\,dy & ; y \geq 0 \\ 0\,dy & ; \text{ otherwise} \end{cases} \tag{9.18}$$

The probability of detection is given by

$$P_D = \int_\eta^\infty dy \, p_{Rice}(y)$$

$$= 1 - \mathrm{CDF}_{Rice}(\eta)$$

$$= Q_1\left(\|a\sqrt{n_s}\|\sqrt{2}, \eta\sqrt{2}\right), \tag{9.19}$$

where $Q_M(v, \mu)$ is the Marcum Q-function that is discussed in Section 13.10.6 for $M = 1$. In Figure 9.2, we display the ROC for the detection test statistic under the assumption of per-sample SNR of 0 dB for 10, 4, and 1 samples. It is worth noting that the performance is noticeably superior to that of the energy detector that we depicted in Figure 9.1.

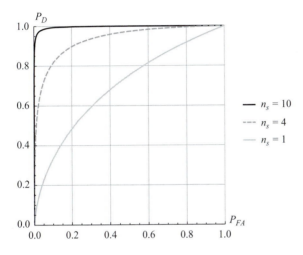

Figure 9.2 Receiver operating characteristic performance of cross-correlation detector for SNR of 0 dB. Single-, 4-, and 10-sample performance is compared.

9.2.3 Normalized Inner Product

An important extension to the cross-correlation detector is the normalized inner product of the n_s samples. We often do not know the variance of the interference-plus-noise environment. This is particularly true when the training or pilot signal is present but misaligned. Consequently, a form that reduces the sensitivity to this effect is useful. We expect this form to have a slightly worse ROC but to be more robust to model errors. The normalized inner product provides such a detection test statistic. It is given by

$$\phi = \frac{\|\mathbf{z}\,\mathbf{s}^H\|^2}{\|\mathbf{z}\|^2\|\mathbf{s}\|^2}. \tag{9.20}$$

The values of this detection test statistic ϕ are bounded by 0 and 1. We can expand this detection test statistic, which gives us

$$\phi = \frac{\|\mathbf{z}\,\underline{\mathbf{s}}^H\|^2}{\|\mathbf{z}\|^2\,\|\underline{\mathbf{s}}\|^2}$$

$$= \frac{\mathbf{z}\,\underline{\mathbf{s}}^H\,\underline{\mathbf{s}}\,\mathbf{z}^H}{\|\mathbf{z}\|^2\,\|\underline{\mathbf{s}}\|^2}$$

$$= \frac{\mathbf{z}\,\underline{\mathbf{s}}^H\,(\underline{\mathbf{s}}\,\underline{\mathbf{s}}^H)^{-1}\,\underline{\mathbf{s}}\,\mathbf{z}^H}{\|\mathbf{z}\|^2}$$

$$= \frac{\mathbf{z}\,\mathbf{P}_{\underline{\mathbf{s}}}\,\mathbf{z}^H}{\|\mathbf{z}\|^2}. \tag{9.21}$$

Under the null hypothesis $\mathcal{H}_0$, there is only noise, so

$$\phi = \frac{\mathbf{n}\,\mathbf{P}_{\underline{\mathbf{s}}}\,\mathbf{n}^H}{\|\mathbf{n}\|^2}, \tag{9.22}$$

where $\underline{\mathbf{n}} \in \mathbb{C}^{1 \times n_s}$ is drawn from a zero-mean circularly symmetric complex Gaussian distribution. The denominator is given by the χ^2 distribution with $2\,n_s$ real dimensions (2 because of complex values). The projection operator takes the $2\,n_s$ real dimensions and projects them onto a single dimension of the same variance. Thus, this operation produces a random variable drawn from a χ^2 distribution with a single degree of freedom. This projection is one component of the original vector $\underline{\mathbf{n}}$. This ratio is given by the central $\beta(a,b)$ distribution [19]. The random variable ϕ under the $\mathcal{H}_0$ hypothesis is given by

$$\phi \sim \beta\left(\frac{1}{2}, \frac{n_s - 1}{2}\right). \tag{9.23}$$

The technical details of these calculations are beyond the scope of this text, but we will provide the function forms of the probabilities of detection and false alarm. The probability of false alarm is given by 1 minus the CDF for this distribution $P_\beta(\eta; n_s)$, and is given by

$$P_{FA}(\eta) = 1 - P_\beta(\eta; n_s) = 1 - \frac{B[\eta; 1/2, (n_s - 1)/2]}{B[1/2, (n_s - 1)/2]}; \quad 0 < \eta < 1, \tag{9.24}$$

where $B[a, b]$ is the β function and $B[x; a, b]$ is the incomplete β function (which should not be confused with the β distribution) [19].

Under the hypothesis that the signal is present and aligned with the reference, the detection test statistic ϕ is given by

$$\phi = \frac{\mathbf{z}\,\mathbf{P}_{\underline{\mathbf{s}}}\,\mathbf{z}^H}{\|\mathbf{z}\|^2}$$

$$= \frac{\|[a\,\underline{\mathbf{s}} + \underline{\mathbf{n}}]\,\mathbf{P}_{\underline{\mathbf{s}}}\|^2}{\|a\,\underline{\mathbf{s}} + \underline{\mathbf{n}}\|^2}$$

$$= \frac{\|a\,\underline{\mathbf{s}} + \underline{\mathbf{n}}\,\mathbf{P}_{\underline{\mathbf{s}}}\|^2}{\|a\,\underline{\mathbf{s}} + \underline{\mathbf{n}}\|^2}. \tag{9.25}$$

In this case, the detection test statistic is given by the noncentral $\beta(a, b, \gamma)$ distribution, where γ indicates the noncentrality defined by the noncentrality parameter of the χ^2 distribution. The detection test statistic ϕ is drawn from

$$\phi \sim \beta\left(\frac{1}{2}, \frac{n_s - 1}{2}, \gamma\right) . \tag{9.26}$$

The noncentrality parameter is proportional to $\|a\underline{s}\|^2$, but we have to consider the normalization. The underlying assumption for the standard χ^2 distribution is that it is constructed from univariance real Gaussian random variables. For our complex data, we can indicate the noise variance σ^2, which is constructed from two Gaussian variables. The variance for the real Gaussian is then given by $\sigma^2/2$. The noncentrality parameter γ is consequently given by

$$\gamma = \frac{\|a\underline{s}\|^2}{\sigma^2/2} = \frac{2\|a\underline{s}\|^2}{\sigma^2} = 2\,\text{ISNR}, \tag{9.27}$$

where ISNR is the integrated SNR $= n_s$ SNR. The probability of detection is given by

$$P_D(\eta) = 1 - \sum_{m=0}^{\infty} \frac{1}{m!} \left(\frac{\gamma}{2}\right)^m e^{-\gamma/2} \frac{B[\eta; 1/2 + m, (n_s - 1)/2]}{B[1/2 + m, (n_s - 1)/2]} ; \quad 0 < \eta < 1. \tag{9.28}$$

We compare the performance of the ROC with the performance of the cross-correlation and normalized inner product test statistics in Figure 9.3. For the comparison, we assume an SNR of 0 dB and 10 independent samples. Because the curves are relatively close on a linear plot, we use a semilog presentation. The cross-correlation approach performs noticeably better in the region of very small probabilities of false alarms. This is the important region because the communications system will be constantly testing for acquisition, and anything but the smallest of false alarm rates would overwhelm the radio system. However, the normalized

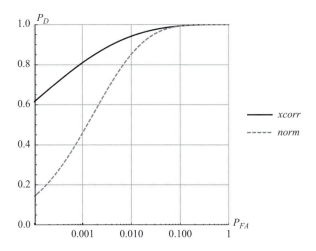

Figure 9.3 Receiver operating characteristic performance of cross-correlation "xcorr" and normalized inner product "norm" test statistics are compared under the assumption of SNR of 0 dB and 10 independent samples.

inner product test statistic is robust to uncertainty in the noise floor. Conversely, small errors in the estimate of the noise variance could cause significant errors in the false alarm rates for the test statistic. For a practical system, one would operate at a higher ISNR, which would allow the system to operate with improved ROC performance. Consequently, a higher probability of detection could be maintained while significantly reducing the probability of false alarm.

9.2.4   Autocovariance Detector

There are various signal detectors that take advantage of known waveform characteristics, such as those that exploit changes in higher-order cumulants of signals. Here, we will consider exploiting the cyclic prefix that orthogonal frequency-division multiplexing (OFDM) modulations typically employ, as was discussed in Section 7.5.2.

Because the length of the OFDM symbol and the length of the cyclic prefix are known at the receiver, this information can be exploited to build an acquisition approach that is independent of the details of the signal. In particular, for an OFDM symbol of length n_s samples with cyclic prefix of length n_{cyc} samples, an autocovariance detection test statistic has the form

$$\phi = \left\| \sum_{m=1}^{n_{cyc}} z_m z_{m+n_s}^* \right\|^2 , \tag{9.29}$$

where z_m indicates the observed samples. We assume here that z_m is zero mean or has had its mean removed. Signal acquisition is declared if ϕ exceeds some threshold. While we have previously focused on just the two possible hypotheses, in general there are a number of potential hypotheses to consider:

- H_0 – just thermal noise;
- H_1 – signal perfectly aligned within sample region;
- H_2 – partially aligned within the sample region; and
- H_3 – some other signal that does not satisfy the expected OFDM structure.

A complete analysis would consider all of these hypotheses.

Similar to the normalized inner product in Section 9.2.3, we can produce a more robust version of this test statistic by normalizing the input data, which is given by

$$\tilde{\phi} = \frac{\left\| \sum_{m=1}^{n_{cyc}} z_m z_{m+n_s}^* \right\|^2}{\sum_{m=1}^{n_{cyc}} \|z_m\|^2 \sum_{m=1}^{n_{cyc}} \|z_{m+n_s}\|^2} . \tag{9.30}$$

9.3   Frequency Synchronization

Because the transmitter and receiver have independent local oscillators or physical Doppler-frequency shifts, they disagree about carrier frequency f_c. The frequency error Δf may be slight, from $\delta f \approx 10^{-6} f_c$ to $10^{-8} f_c$, but the phase error can be sufficient to cause a decoding error, given a sufficiently long period of observation. Consequently, the communications system typically attempts to correct this error. This correction is sometimes called *carrier recovery* or *frequency synchronization*. It is worth noting that the model that we suggest here is a bit simplistic. For real environments, there is spread of frequency errors, which may be caused by the phase noise in the local oscillators or by multiple scatterers in the environment that introduce different Doppler-frequency shifts (or Doppler-frequency spread). However, a single frequency shift model is often sufficient, or at least a good starting point, so we assume this model here.

Historically, analog communications systems would exploit the strong carrier combined with a phase-lock loop (PLL) to look onto the carrier frequency. A Costas loop, which integrates PLL with in-phase and quadrature channels, can be used to recover the carrier for digital modulations. With modern communications systems that have reasonable computational capability, we can employ metrics digitally evaluated at complex baseband. In particular, most digital communications systems employ training or pilot sequences. We can determine the frequency error by comparing the phase progression between the observed signal and the reference signal.

Here, we introduce a maximum likelihood estimator for frequency synchronization. We consider a single-carrier flat-fading model with a sequence n_s of known transmitted training symbols arranged into a column vector $\mathbf{s} \in \mathbb{C}^{n_s \times 1}$. The complex channel attenuation is represented by a, and complex additive Gaussian noise is denoted by $\mathbf{n} \in \mathbb{C}^{n_s \times 1}$. We assume that, from the receiver's perspective, the transmit sequence is distorted by a frequency shift δf so that

$$s_m \rightarrow s_m \, e^{i 2\pi \, \delta f \, T \, (m-1)} , \tag{9.31}$$

where T is the sample period. The observed samples at the receiver are denoted by $\mathbf{z} \in \mathbb{C}^{n_s \times 1}$ and are given by

$$\mathbf{z} = a \, \mathbf{s} \odot \mathbf{f} + \mathbf{n} . \tag{9.32}$$

We will develop the maximum likelihood estimator, which is given by

$$\hat{\delta f} = \operatorname{argmax}_{\delta f, a} \log_e p(\mathbf{z}; a, \delta f) . \tag{9.33}$$

The general form of the likelihood (probability density function) for this model is given by

$$p(\mathbf{z}; a, \delta f) = \frac{1}{\pi^{n_s} |\mathbf{R}|} e^{-(\mathbf{z} - a \, [\mathbf{s} \odot \mathbf{f}])^H \, \mathbf{R}^{-1} \, (\mathbf{z} - a \, [\mathbf{s} \odot \mathbf{f}])} , \tag{9.34}$$

where $\mathbf{R}$ is the covariance of noise terms such that the lth, kth element of the matrix is given by $\mathbf{R}_{l,k} = E\{n_l \, n_k^*\}$. We can assemble the effects of the frequency shift into the vector $\mathbf{f}$, given by

$$f = \begin{pmatrix} e^{i\, 2\pi\, \delta f\, T\, 0} \\ e^{i\, 2\pi\, \delta f\, T\, 1} \\ \vdots \\ e^{i\, 2\pi\, \delta f\, T\, (n_s - 1)} \end{pmatrix}. \tag{9.35}$$

As an example, if $\delta f = 0$, then this vector is just the all-1s vector and has no effect on **s**.

In many cases the maximization of the likelihood must be performed by explicitly trying all values of the parameters. However, in this case, we are fortunate in that an algebraic solution can be found for the channel coefficient. Because a is a complex parameter, we optimize by setting the derivative with respect to some real parameter α to zero, as discussed in Section 13.8. By using this approach, the maximum likelihood solution is given by

$$a = a(\alpha) \tag{9.36}$$

$$\frac{\partial}{\partial \alpha} \log_e p(z) = \left(\frac{\partial}{\partial \alpha} a^* \right) [-a\, (s \odot f)^H\, \mathbf{R}^{-1}\, (s \odot f) + (s \odot f)^H\, \mathbf{R}^{-1}\, z] + c.c. \tag{9.37}$$

$$\Rightarrow a\, (s \odot f)^H\, \mathbf{R}^{-1}\, (s \odot f) = (s \odot f)^H\, \mathbf{R}^{-1}\, z \tag{9.38}$$

$$\hat{a} = \frac{(s \odot f)^H\, \mathbf{R}^{-1}\, z}{(s \odot f)^H\, \mathbf{R}^{-1}\, (s \odot f)}. \tag{9.39}$$

We can then substitute the estimate of the channel coefficient into the likelihood and simplify:

$$\log_e p(z; \hat{a}, \delta f) = k - z^H\, \mathbf{R}^{-1}\, z$$

$$- \frac{(s \odot f)^H\, \mathbf{R}^{-1}\, z}{(s \odot f)^H\, \mathbf{R}^{-1}\, (s \odot f)} \left[\frac{(s \odot f)^H\, \mathbf{R}^{-1}\, z}{(s \odot f)^H\, \mathbf{R}^{-1}\, (s \odot f)} \right]^* (s \odot f)^H\, \mathbf{R}^{-1}\, (s \odot f)$$

$$+ \frac{(s \odot f)^H\, \mathbf{R}^{-1}\, z}{(s \odot f)^H\, \mathbf{R}^{-1}\, (s \odot f)}\, z^H\, \mathbf{R}^{-1}\, (s \odot f)$$

$$+ \left[\frac{(s \odot f)^H\, \mathbf{R}^{-1}\, z}{(s \odot f)^H\, \mathbf{R}^{-1}\, (s \odot f)} \right]^* (s \odot f)^H\, \mathbf{R}^{-1}\, z$$

$$= k - z^H\, \mathbf{R}^{-1}\, z + \frac{\left[(s \odot f)^H\, \mathbf{R}^{-1}\, z \right]^* (s \odot f)^H\, \mathbf{R}^{-1}\, z}{(s \odot f)^H\, \mathbf{R}^{-1}\, (s \odot f)}. \tag{9.40}$$

Unfortunately, an algebraic solution for δf does not exist. The maximum likelihood solution is performed by searching over a set of possible frequency offset values and is denoted as

$$\hat{\delta f} = \operatorname{argmax}_{\delta f} \log_e p(z; \hat{a})$$

$$= \operatorname{argmax}_{\delta f} \frac{\left[(s \odot f)^H\, \mathbf{R}^{-1}\, z \right]^* (s \odot f)^H\, \mathbf{R}^{-1}\, z}{(s \odot f)^H\, \mathbf{R}^{-1}\, (s \odot f)}$$

$$= \operatorname{argmax}_{\delta f} \frac{\left\| (s \odot f)^H\, \mathbf{R}^{-1}\, z \right\|^2}{\left\| \mathbf{R}^{-1/2}\, (s \odot f) \right\|^2}. \tag{9.41}$$

If we assume that the noise is independently and identically distributed, then the noise covariance matrix is replaced with something that is proportional to the identity matrix: $\mathbf{R} \rightarrow \sigma_n^2 \mathbf{I}$. The determinant is then given by $|\sigma_n^2 \mathbf{I}| = \sigma_n^{2 n_s}$.

$$\hat{\delta f} = \mathrm{argmax}_{\delta f} \log_e p(z; \hat{a}, \delta f) \tag{9.42}$$

$$= \mathrm{argmax}_{\delta f} \sigma_n^{-2 n_s} \frac{\left\| (\mathbf{s} \odot \mathbf{f})^H \mathbf{z} \right\|^2}{\| \mathbf{s} \odot \mathbf{f} \|^2}$$

$$= \mathrm{argmax}_{\delta f} \left\| (\mathbf{s} \odot \mathbf{f})^H \mathbf{z} \right\|^2 . \tag{9.43}$$

It is interesting to note that the result is given by maximizing the vector inner product between the model of the transmitted signal and the vector of observed data.

9.4 Fine-Time Synchronization

In a process similar to that for achieving frequency synchronization, we can employ a known training or pilot sequence to precisely determine timing offsets. For this discussion, we assume a single-carrier system that is sending training sequence samples represented by the vector $\mathbf{s} \in \mathbb{C}^{n_s \times 1}$. We assume a shifted version of this sequence by some delay τ is indicated by $\mathbf{s}_\tau$. The model that we have is given by

$$p(\mathbf{z}; a, \delta f) = \frac{1}{\pi^{n_s} |\mathbf{R}|} e^{-(\mathbf{z} - a \mathbf{s}_\tau)^H \mathbf{R}^{-1} (\mathbf{z} - a \mathbf{s}_\tau)}, \tag{9.44}$$

where $\mathbf{R}$ is the covariance of noise terms such that the lth, kth element of the matrix is given by $\mathbf{R}_{l,k} = E\{n_l n_k^*\}$. Similar to the frequency shift estimator in Equation (9.41), the delay estimation is given by

$$\hat{\tau} = \mathrm{argmax}_\tau \log_e p(\mathbf{z}; \hat{a})$$

$$= \mathrm{argmax}_\tau \frac{\left[\mathbf{s}_\tau^H \mathbf{R}^{-1} \mathbf{z} \right]^* \mathbf{s}_\tau^H \mathbf{R}^{-1} \mathbf{z}}{\mathbf{s}_\tau^H \mathbf{R}^{-1} \mathbf{s}_\tau}$$

$$= \mathrm{argmax}_\tau \frac{\left\| \mathbf{s}_\tau^H \mathbf{R}^{-1} \mathbf{z} \right\|^2}{\| \mathbf{R}^{-1/2} \mathbf{s}_\tau \|^2}, \tag{9.45}$$

which reduces to

$$\hat{\tau} = \mathrm{argmax}_\tau \left\| \mathbf{s}_\tau^H \mathbf{z} \right\|^2 \tag{9.46}$$

if the noise is uncorrelated from sample to sample.

9.4.1 Spectral-Domain Timing Estimation

Another timing estimation approach for OFDM waveforms is to exploit the Fourier relationship given by

$$\mathcal{F}\{s(t)\} = S(f)$$

$$\mathcal{F}\{s(t - \tau)\} = Z(f) = e^{i 2\pi f t} S(f). \tag{9.47}$$

Here, we assume that we have captured approximately the right range of data for $s(t)$, and we are trying to find correction. For the OFDM waveform, if we use a subset of carriers as training or pilot data, then we know what the spectrum should be for some set of baseband frequencies. Consequently, we know the expected spectral value at some set of frequencies f_m. We define the expected spectral values to be

$$S(f_m) \forall f_m \in \mathcal{T}, \tag{9.48}$$

where $\mathcal{T}$ is the set of discrete training frequencies. We can now compare the expected spectral responses to the actual ones by multiplying by the conjugate of the reference at each frequency. Furthermore, the delay information is given by the phase ramp across frequencies. We measure the relative phase at each frequency:

$$\phi_m \angle [Z(f_m) \, S^*(f_m)], \tag{9.49}$$

where ϕ_m is the phase of the inner product for subcarrier m. The delay is related to the phase ramp by

$$\phi_m = 2\pi f_m \tau. \tag{9.50}$$

We can unwrap this phase to make a straight line and evaluate the slope to find τ.

The final challenge of this approach is that the channel is typically dispersive and has noise. Consequently, we get noisy measurements. We can use a least-squares estimator to fit to the noisy data.

Problems

9.1 For a radio link that operates at a carrier frequency of near 1 GHz, evaluate the frequency offsets contributions if the closing velocity is 10 m/s and the relative local oscillator is given by two parts in a million.

9.2 In simulation, evaluate the probability of detection and probability of false alarm for an energy detector. For each sample, assume complex Gaussian signals and unit-variance noise with a signal SNR of 6 dB. Set the detection threshold $q \gtrless \eta$ at $\eta = 3$ dB. Use the metric

$$q = \frac{1}{N} \sum_m^N \|x_m\|^2.$$

Estimate P_D and P_{FA} for
(a) $N = 1$
(b) $N = 4$
(c) $N = 16$
(d) $N = 64$
To estimate P_D and P_{FA} for each value of N, repeat by simulating 1000 trials.

9.3 Evaluate the ROC curve for the following no-information test statistic. If the signal is absent, the q is drawn from a uniform distribution from 0 to 10. If the signal is present, the q is drawn from a uniform distribution from 0 to 10. Evaluate the probability of detection and false alarm, and plot the ROC curve.

9.4 By using simulation, employ an energy detection approach for a pseudorandom QPSK signal that has a length of 10 chips and the received SNR of 0 dB with noise variance of σ_n^2. Evaluate the probability of detection and false alarm for the following threshold values:

(a) $\eta = 10\,\sigma_n^2$
(b) $\eta = 20\,\sigma_n^2$
(c) $\eta = 30\,\sigma_n^2$
(d) $\eta = 40\,\sigma_n^2$

9.5 Compare the simulated performance in Problem 9.4 to the theoretical ROCs for the following threshold values:

(a) $\eta = 10\,\sigma_n^2$
(b) $\eta = 20\,\sigma_n^2$
(c) $\eta = 30\,\sigma_n^2$
(d) $\eta = 40\,\sigma_n^2$

9.6 Collect the probability of detection and false alarm for the normalized inner product detector.

9.7 For a pilot sequence that is 16 samples and has a received SNR of -1 dB, mathematically evaluate the ROC performance on a plot with linear probability of detection and logarithmic probability of false alarm (10^{-6} to 1) for

(a) energy detection;
(b) cross-correlation; and
(c) normalized inner product approaches.

9.8 By using simulation, consider a normalized inner-product approach for detection for a 10-chip pseudorandom QPSK signal that has a received average SNR of 3 dB for each chip. Find the frequency offset such that the probability of detection is reduced to half the probability without the frequency offset.

9.9 Consider N samples of a single-carrier flat-fading model of a communications system with frequency offset δf, which is given by

$$\mathbf{z} = a\,\mathbf{s} \odot \mathbf{f} + \mathbf{n}.$$

(a) Construct the vector $\mathbf{f}$.
(b) By assuming that the noise is complex, identical, and independent, present the probability density function for $\mathbf{z}$.
(c) Find the maximum likelihood estimator for a.
(d) Show that the maximum likelihood estimator for δf is given by

$$\hat{\delta f} = \text{argmax}_{\delta f} \left\| (\mathbf{s} \odot \mathbf{f})^H \mathbf{z} \right\|.$$

9.10 By using simulation, consider the autocovariance detection performance of an OFDM symbol with a cyclic prefix of length 10 chips. Find the required SNR to achieve a probability of detection of greater than 0.95 and a probability of false alarm of less than 10^{-4}.

9.11 By using simulation, develop a 128-subcarrier OFDM symbol in which every fourth subcarrier is a training symbol that is a known QPSK symbol. Consider a received signal in noise. By assuming a flat-fading channel and an SNR of 6 dB, construct a received signal for which the received signal is delayed by two samples in time. Plot the measured phase ramp by using the training data. Search over possible fractional delays to find the best match to the observed phase ramp.

10 Radio Duplex, Access, and Networks

In this chapter we discuss how multiple radios interact. We introduce the concept of duplex and various approaches to enable radios to perform bidirectional communications. We also introduce the concept of network topologies such as star and mesh approaches. We discuss multiple media access control techniques. We introduce aloha, carrier-sense multiple access, time-division multiple access, frequency-division multiple access, and code-division multiple access.

10.1 Duplexing Approaches

We have presented communications as a one-way process, from a transmitter to a receiver. However, often communications nodes are transceivers, which can operate as both transmitters and receivers. In general, some deconfliction approach must be employed. There are various ways that two nodes may organize the exchange of information. Arguably, the most intuitive is time-division duplex (TDD) communications, which is how humans communicate: each individual taking turns (unless we are talking over each other).

Another common approach is frequency-division duplex (FDD) communications. With this approach, each link direction is given its own frequency to use. While, theoretically, this seems easy, if the two frequencies are close together, leakage through the spectral sidelobes from the transmitted signal into the radio receiver can cause overwhelming self-interference.

One last approach is in-band full-duplex (IBFD) communications [20], which is sometimes just called *full-duplex*, although this nomenclature can lead to confusion between IBFD and FDD. This is a particularly challenging approach because each radio transmits and receives at the same time on the same frequency. Sometimes, IBFD includes using the same antenna for both transmission and reception. The technical challenge here is that the self-interference must be suppressed with exquisite precision.

10.2 Multiple Links

Up to this point we have generally assumed that only one communications link exists in the universe. As you might imagine, this assumption is a bit simplistic. The density of potential users varies significantly as one moves from rural to urban areas. It can even change significantly over time – for example, because of diurnal variation or popular events. If we have finite resources, we need to decide how to allocate these resources to various links. We call the process of making and implementing these decisions *media access control* (MAC), as was introduced in Section 2.3.2.

We often call the connection between radio nodes a *network*. In general, there are an infinite number of network topologies (that is, how the nodes are connected by communication links). However, we can capture many important ideas with just a few topologies. A simple one is the broadcast network. In this case, a single node transmits the same message to a group of nodes. Unsurprisingly, broadcast television and radio are examples of this topology. Similar to this approach is the star topology, which has a central node, base station, or access point with which a group of radios communicate. An example is depicted in Figure 10.1(a). In this case, the base station communicates to each node individually, and the mobile nodes communicate only to the base station. Another important topology is the mesh network, in which nearby nodes communicate with each other. An example is depicted in Figure 10.1(b). This network might be fully connected, which is to say every node can communicate with every other node. However, often only closer links can communicate. In this case, for any two nodes to communicate, multiple "hops" are required, so a message is passed from source to node to node until it reaches its destination. For either the star or mesh topology, some mechanism to allocate the limited resources must be determined. Approaches to allocating these resources often try to impose some sort of fairness or to allocate more resources to nodes with greater need.

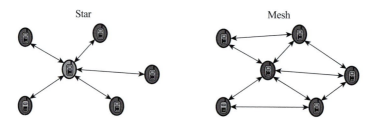

Star   Mesh

Figure 10.1 Network topologies: (a) star network, (b) mesh network.

10.3   Media Access Control

10.3.1   Aloha and Carrier-Sense Multiple Access

For the sake of argument, let us assume that there is some frequency allocation that a group of radios are allowed to use. The first approach that one might consider is to just transmit when you have something to communicate. This approach is called *aloha access*. For this access approach, an acknowledgment message is returned when a message is received. If the original transmitter does not receive an acknowledgment, it resends the message. For a network of nodes, this approach will only work when the channel usage is very low. If the rate of messages that are being sent is high, then the probability of message collision becomes large, which causes more messages to be resent, which causes more collisions, and the system performance collapses.

A slight modification of this approach is carrier-sense multiple access (CSMA). For CSMA, the radio senses the channel in an attempt to detect another communications signal. If it is determined that the channel is in use, then after some random delay, the channel is checked again. The clear advantage of this approach is that it is relatively simple to implement. On the down side, small amounts of accidental interference can cause the channel to be unused. It is also possible that the channel is being used but the carrier-sense algorithm does not detect it because of sensitivity or geometry. In this case, despite the radio's best attempt to be a good neighbor, it causes another link to fail. This concern is sometimes called the *hidden-node problem*. An example of a geometry that could potentially cause this effect is depicted in Figure 10.2. The carrier-sensing transmitter is relatively close to a neighboring-link receiver but is far from the neighboring-link transmitter and, consequently, does not detect its presence.

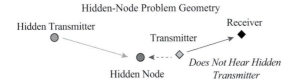

Figure 10.2 Geometry that may cause the hidden node to be missed by carrier-sensing approaches.

10.3.2 Time-Division Multiple Access

Another important approach for enabling multiple access to the channel is time-division multiple access (TDMA). This approach allocates each user time slots for the channel. A simple example is depicted in Figure 10.3. Here, eight users are given one out of eight slots that repeat. Potentially, the slot allocation can be more complicated. To enable this approach, some mechanism must be provided to assign slots.

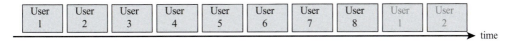

Figure 10.3 Simple example of time slot allocation for the TDMA approach.

10.3.3 Frequency-Division Multiple Access

The frequency-division multiple access (FDMA) approach (Figure 10.4) allocates different frequencies for different communications links. The concept of FDD, discussed in Section 10.1, took this even further in that different communications directions were given different frequencies. Implicitly, we have been assuming the concept of FDMA in a coarse sense. The idea for FDMA here is that, for some network, different links are given different frequencies within some larger allocation given to the overall network. Similar to the organizational issues associated with TDMA, some mechanism must assign frequencies.

Frequency

Figure 10.4 Simple example of spectral allocation for the FDMA approach.

10.3.4   Code-Division Multiple Access

Most multiple-access approaches assume that communications approaches use different links that are isolated in frequency or time. The code-division multiple access (CDMA) approach removes this constraint by separating signals in code. CDMA networks typically operate with relatively low spectral efficiency for a given link. The symbols associated with each radio are "spread" with a known pseudorandom code. A matched filter is used at the receiver to "de-spread" the signal and recover the signal of interest while suppressing interfering signals. Ideally, these spreading codes are orthogonal so that interference from other users can be completely mitigated with the matched filter. In practice, channel delay and Doppler spread break orthogonality. Consequently, it is often important to control the power of the received signals so that they are approximately equal.

We consider a simple CDMA example here. To aid understanding, we keep the dimensionality relatively small. We consider an example with two users transmitting simultaneously to a receiver. We assume that the users are employing a binary-phase-shift keying (BPSK) modulation. The users' transmissions are synchronous in time and occupy the same spectrum. We assume simple length-4 Walsh or Hadamard codes and select two of the orthogonal sequences for spreading. The spreading codes for users 1 and 2 are

$$\begin{aligned} \underline{c}_1 &= (1 \quad 1 \quad 1 \quad 1) \\ \underline{c}_2 &= (1 \quad 1 \quad -1 \quad -1) \end{aligned}. \tag{10.1}$$

If users 1 and 2 have sequences

$$\begin{aligned} \underline{s}_1 &= (1 \quad 1 \quad -1) \\ \underline{s}_2 &= (-1 \quad 1 \quad 1) \end{aligned} \tag{10.2}$$

to send, then the spread sequences are given by

$$\begin{aligned} \tilde{\underline{s}}_1 &= \underline{s}_1 \otimes \underline{c}_1 = (1 \quad 1 \quad 1 \quad 1 \quad 1 \quad 1 \quad 1 \quad 1 \quad -1 \quad -1 \quad -1 \quad -1) \\ \tilde{\underline{s}}_2 &= \underline{s}_2 \otimes \underline{c}_2 = (-1 \quad -1 \quad 1 \quad 1 \quad 1 \quad 1 \quad -1 \quad -1 \quad 1 \quad 1 \quad -1 \quad -1) \end{aligned}, \tag{10.3}$$

where $\otimes$ is the Kronecker product. The received signal $\underline{z}$ is then given by

$$\underline{z} = a_1 \tilde{\underline{s}}_1 + a_2 \tilde{\underline{s}}_2 + \underline{n}, \tag{10.4}$$

where a_1 and a_2 are channel coefficients and $\underline{n}$ is a vector of Gaussian noise. For the sake of discussion and simplifying the explanation, let us assume that $a_1 = a_2 = 1$. In practice, we can adaptively compensate for the channel. Also, let us ignore the effects of the noise for the moment. Consequently, the received signal is approximately

$$\underline{z} \approx \tilde{\underline{s}}_1 + \tilde{\underline{s}}_2. \tag{10.5}$$

We can de-spread the received signal by first remapping its shape into a matrix $\mathbf{Z}$ that has a number of rows equal to the length of the spreading codes:

$$\mathbf{Z} = \begin{pmatrix} z_1 & z_5 & z_9 \\ z_2 & z_6 & z_{10} \\ z_3 & z_7 & z_{11} \\ z_4 & z_8 & z_{12} \end{pmatrix}. \tag{10.6}$$

The de-spread signals for $\hat{\underline{s}}_1$ and $\hat{\underline{s}}_2$ are then given by

$$\begin{aligned} \hat{\underline{s}}_1 = \underline{c}_1^* \mathbf{Z} &= 4\,(1 \quad 1 \quad -1) \\ \hat{\underline{s}}_2 = \underline{c}_2^* \mathbf{Z} &= 4\,(-1 \quad 1 \quad 1) \end{aligned}. \tag{10.7}$$

From the construction of the de-spreading approach, one can observe that the length of the spreading sequence must be longer than the number of users in the environment. In general, adaptive extensions to the simple de-spreading approach presented here need to be employed to reliably recover the signals in dispersive environments.

In the presence of noise, de-spreading effectively increases the signal-to-noise ratio (SNR) by the spreading length. The SNR per symbol is increased compared to the per-sample SNR. This is because the de-spreading combines the spread samples coherently, which produces a spreading length squared single power increase, while the noise combines incoherently, so the noise power increases linearly.

For CDMA, we can imagine fixing the source symbol rate, or fixing the output symbol rate, depending upon one's view. In the former case, the chip rate, and consequently the bandwidth of the transmitted signal, is increased by approximately a factor equal to the length of the spreading code. In the latter case, the bandwidth of the transmitted signal is held constant, but the data rate is reduced by a factor equal to the length of the spreading code.

Problems

10.1 Identify the advantages and disadvantages of using TDD versus FDD.

10.2 Assume a scenario with three nodes that jointly use a common 1 MHz bandwidth, 0 dBi antennas, temperature of 300 K, and a carrier frequency of 1 GHz. By assuming that all the nodes lie along a line, line-of-sight propagation, and

1 mW transmit powers, construct an example of the hidden-node problem. There are multiple ways to construct this problem.

10.3 For a network with eight mobile stations and one base station, in which the average data rate is 1 Mb/s for each node, with a bandwidth of 16 MHz, construct example waveforms for the uplink for
(a) TDMA
(b) FDMA

10.4 Consider a CDMA system with two BPSK transmitters, each sending 1 Mb/s to a single receiver. By assuming 10 MHz bandwidth and a flat-fading channel, design BPSK spreading sequences. By hand evaluation, demonstrate the receiver's ability to disentangle the transmitted symbols.

10.5 Consider a CDMA system with eight QPSK transmitters and QPSK spreading sequences of length 32 symbols. Design the spreading sequences. By using a simulation of the flat-fading Gaussian noise channel, evaluate and plot the de-spread hard decision symbol error rates for all eight transmitters by averaging over 1000 trials for per received sample SNR of -10 dB to 20 dB in 1-dB steps.

11 Multiple-Antenna and Multiple-Input Multiple-Output Communications

In this chapter we discuss the use of multiple antennas by radios. While the radio links that we discussed up to this point in the text have assumed single-input single-output (SISO) channels, we now consider the use of multiple antennas at both the transmitter (source) and receiver (destination), as indicated in Figure 11.1. We introduce the channel model for a multiple-antenna receiver. We discuss channel estimation and spatial receive beamforming techniques. We introduce the multiple-input multiple-output (MIMO) channel model, define the capacity of this system under the assumptions that the transmitter is uninformed and the transmitter is informed of the channel matrix. Finally, we discuss the concept of space–time coding and present various approaches, including Alamouti's space–time block code.

For the discussion within this chapter, we will assume that the channels are spectrally flat (not frequency-selective). There are frequency-selective channel extensions to the discussion provided here [21].

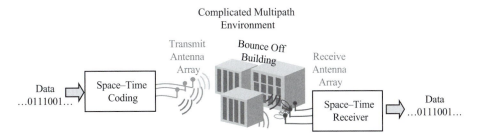

Figure 11.1 Notional 3 × 3 MIMO communications link.

11.1 Multiple-Antenna Receivers

We consider a scenario in which there is a transmitter with a single antenna and a receiver with multiple antennas. This is a single-input multiple-output (SIMO) channel. We assume that there is multipath scattering but that the delay spread of the scattering is not easily resolved because the temporal spread of the arriving delays is small compared to the inverse of the signal bandwidth.

Because there are multiple n_r receive antennas, we can review the received signal (at complex baseband) to be a vector $\mathbf{z}(t) \in \mathbb{C}^{n_r \times 1}$ as a function of time t, which is given by

$$\mathbf{z}(t) = \mathbf{h}\, s(t) + \mathbf{n}(t). \tag{11.1}$$

The complex channel attenuation from the transmitter to each of the receiver antennas for this flat-fading scenario is represented by the vector $\mathbf{h} \in \mathbb{C}^{n_r \times 1}$. The transmitter sends the complex scalar signal $s(t)$. Versions of this signal are received with different amplitudes and phases at each receive antenna. These amplitudes and phases are determined by how the signal bounces through the environment. They are relatively easy to predict in a line-of-sight environment if we assume knowledge of the exact positions and distances between all of the antennas. In the case of complicated scattering environments, which are common in terrestrial communications links, it is nearly impossible to accurately predict the values in $\mathbf{h}$. However, we can estimate them by using training data. The n_r receive antennas have independent receive low-noise amplifiers (LNAs) and mixers. Consequently, each receive antenna has independent noise, and we need a vector where the entries are drawn independently, from antenna to antenna, from a complex circularly symmetric Gaussian distribution: $\mathbf{n}(t) \in \mathbb{C}^{n_r \times 1}$. For this discussion, to simplify the forms of the equations, we assume that the units of power are normalized so that noise has unit variance and is independent from antenna to antenna. Consequently, the spatial covariance matrix of the noise is given by

$$E[\mathbf{n}(t)\, \mathbf{n}^H(t)] = \mathbf{I}. \tag{11.2}$$

While we have introduced the channel in terms of continuous time t, in practice we use sampled data. We can express the n_s samples of the receive, transmit, and noise signals as $\mathbf{Z} \in \mathbb{C}^{n_r \times n_s}$, $\underline{\mathbf{s}} \in \mathbb{C}^{1 \times n_s}$, and $\mathbf{N} \in \mathbb{C}^{n_r \times n_s}$, respectively. The model for the received signal is then given by

$$\mathbf{Z} = \mathbf{h}\underline{\mathbf{s}} + \mathbf{N}. \tag{11.3}$$

If there were two different transmitted signals that are received simultaneously, then the received signal becomes

$$\mathbf{Z} = \mathbf{h}_1\, \underline{\mathbf{s}}_1 + \mathbf{h}_2\, \underline{\mathbf{s}}_2 + \mathbf{N}, \tag{11.4}$$

where we use the subscripts 1 and 2 to indicate the signal and channel from the first and second transmitters, respectively. We can extend this formulation to account for an arbitrary number of transmitters by using the form

$$\mathbf{Z} = \sum_m \mathbf{h}_m\, \underline{\mathbf{s}}_m + \mathbf{N}. \tag{11.5}$$

The expected sampled spatial noise covariance is then given by

$$E\left\{ \frac{\mathbf{N}\mathbf{N}^H}{n_s} \right\} = \mathbf{I}, \tag{11.6}$$

where we assume that our units for noise power are set so that the average noise power over the received bandwidth at a given antenna is 1.

We can use the multiple antennas to help recover the individual signals. This setup is a type of multiaccess receiver. As we have discussed previously, it is common to partition the transmitted signal into n_{train} training and n_{data} data symbols so that $\underline{\mathbf{s}}_m = (\underline{\mathbf{s}}_{m,train} \quad \underline{\mathbf{s}}_{m,data})$ for the mth transmitter. We assume that the data and training sequences of each transmitter are independent.

If we are interested in the channel associated with the path from transmitter 1 to the receive array $\mathbf{h}_1$, then the least square channel estimate for the channel vector $\hat{\mathbf{h}}_1$ is given by

$$\hat{\mathbf{h}}_1 = \mathbf{Z}_{train}\, \underline{\mathbf{s}}_{1,train}^{H}\, (\underline{\mathbf{s}}_{1,train}\, \underline{\mathbf{s}}_{1,train}^{H})^{-1}, \tag{11.7}$$

where $\mathbf{Z}_{train} \in \mathbb{C}^{n_r \times n_{train}}$ is the portion of the data associated with the training sequence.

To recover a good estimate of $\underline{\mathbf{s}}_{1,data}$, we can use the training data to help build an adaptive beamformer to maximize the signal while minimizing the effects of the noise and interference. We can imagine multiplying the received signal at each antenna by some complex number and adding the resulting signal together. The question is then how to optimize the coefficients to produce the best estimate. Given this approach, our beamformer is constructed using

$$\hat{\underline{\mathbf{s}}}_{1,data} = \mathbf{w}^{H}\, \mathbf{Z}_{data}, \tag{11.8}$$

where $\mathbf{w}$ is our adaptive beamformer. The mean squared error between the actual data and the output of the beamformer is given by

$$\begin{aligned}
\epsilon^2 &= \frac{\|\hat{\underline{\mathbf{s}}}_{1,data} - \underline{\mathbf{s}}_{1,data}\|^2}{n_{data}} \\
&= \frac{\|\mathbf{w}^{H}\, \mathbf{Z}_{data} - \underline{\mathbf{s}}_{1,data}\|^2}{n_{data}}.
\end{aligned} \tag{11.9}$$

We can approximate this error by using the training data. The approximation to the error is given by

$$\epsilon^2 \approx \tilde{\epsilon}^2 \tag{11.10}$$

$$= \frac{\|\mathbf{w}^{H}\, \mathbf{Z}_{train} - \underline{\mathbf{s}}_{1,train}\|^2}{n_{train}}. \tag{11.11}$$

We calculate $\mathbf{w}$ to minimize the approximation of this error, which gives

$$\frac{\partial \tilde{\epsilon}^2(\alpha)}{\partial \alpha} = 0 = \frac{\partial}{\partial \alpha} \frac{\|\mathbf{w}^{H}(\alpha)\, \mathbf{Z}_{train} - \underline{\mathbf{s}}_{1,train}\|^2}{n_{train}} \tag{11.12}$$

$$\mathbf{w} = (\mathbf{Z}_{train}\, \mathbf{Z}_{train}^{H})^{-1}\, \mathbf{Z}_{train}\, \underline{\mathbf{s}}_{1,train}^{H}, \tag{11.13}$$

where we assume that the beamformer $\mathbf{w}(\alpha)$ is a function of the real parameter α and use the techniques described in Section 13.8. We estimate the transmitted data $\hat{\underline{\mathbf{s}}}_{1,data}$ by applying the adapted beamformer to the received data, which is given by

$$\hat{\underline{\mathbf{s}}}_{1,data} = \mathbf{w}^{H}\, \mathbf{Z}_{data}. \tag{11.14}$$

11.2   MIMO Channel

By using multiple transmit antennas and multiple receive antennas we can potentially improve communications performance compared to single-antenna links. The MIMO channel between transmitter and receiver then has multiple inputs and multiple outputs [21]. With multiple transmit and receive antennas, we have choices on how to use these spatial degrees of freedom. If the transmitter knows the complex attenuation channel matrix between the transmitter and receiver antenna arrays, it can make more sophisticated choices. For the transmitter's space–time coding, we have choices on how to use the multiple transmit antennas. While there is a continuum of choices, two limits of this continuum are full spatial redundancy and spatial multiplexing. For full spatial redundancy, we code in such a way that information is represented at every transmit antenna at some point in time. A very simple example is to take a stream of forward error correction encoded symbols and send each symbol to each antenna on successive chips. The receiver has to disentangle the overlapped signals at the destination radio. The advantage of this approach is that some transmitters will have better channels than others. The assumption is that it is unlikely for all transmit antennas to have bad channels. The other limit of using these antennas is spatial multiplexing. In this limit, encoded symbols are distributed across transmit antennas without redundancy. This maximizes the data rate if one assumes that symbol errors are unlikely.

The best use of these channels varies depending upon the details of the environment. For example, in simple low-SNR flat-fading scenarios, the best option is to send the same signal from each transmit antenna with the appropriate phase to maximize the power at the receiver. This assumes that the transmitter somehow knows the detail of the channel. In rich-scattering low-SNR environments, for scenarios in which the transmitter does not know the channel, the best thing to do is to send independent signals from each antenna with complete coding redundancy. In the high-SNR limit, performance is improved by reducing the redundancy.

By assuming the channel is not frequency-selective, we can characterize it between a radio with an array of n_t transmit antennas and a radio with an array of n_r receive antennas with a complex channel attenuation matrix at complex baseband, which replaces the complex amplitude for a SISO link. This channel matrix $\mathbf{H} \in \mathbb{C}^{n_r \times n_t}$ contains the complex attenuations between each transmit and receive antenna. The complex baseband signal $\mathbf{z}(t) \in \mathbb{C}^{n_r \times 1}$, as a function of time t, that is observed at the receive array is given by

$$\mathbf{z}(t) = \mathbf{H}\,\mathbf{s}(t) + \mathbf{n}(t), \tag{11.15}$$

where the complex baseband signals transmitted from the array of transmit antennas is given by $\mathbf{s}(t) \in \mathbb{C}^{n_t \times 1}$, and the noise observed at the receiver is given by $\mathbf{n}(t) \in \mathbb{C}^{n_r \times 1}$.

While we have introduced the channel in terms of continuous time t, in practice we use sampled data. We can express the n_s samples of the receive, transmit, and noise signals as $\mathbf{Z} \in \mathbb{C}^{n_r \times n_s}$, $\mathbf{S} \in \mathbb{C}^{n_t \times n_s}$, and $\mathbf{N} \in \mathbb{C}^{n_r \times n_s}$, respectively. The model for the received signal is then given by

$$\mathbf{Z} = \mathbf{H}\mathbf{S} + \mathbf{N} . \tag{11.16}$$

11.3 MIMO Capacity

The capacity of the MIMO channel has a similar form to the capacity for a SISO channel, presented in Equation (3.3). Because the channel has a matrix form, it is not surprising that the details must be modified to some extent. The 1 is replaced with the identity matrix $\mathbf{I}$, and there is a determinant of the resulting matrix. The MIMO channel capacity has the form,

$$c = \log_2 |\mathbf{I} + \mathbf{H}\mathbf{P}\mathbf{H}^H| , \tag{11.17}$$

where the channel matrix $\mathbf{H} \in \mathbb{C}^{n_r \times n_t}$ indicates the complex attenuation between transmit and receive antennas, and the matrix $\mathbf{P} \in \mathbb{C}^{n_t \times n_t}$ indicates the noise-normalized transmit covariance matrix:

$$\mathbf{P} = E\{\mathbf{s}(t)\, \mathbf{s}^H(t)\} . \tag{11.18}$$

At first in our discussion, detailed in Section 11.3.1, we assume that the transmitter does not know the channel matrix or complex attenuations between transmit and receive antennas. Then, in Section 11.3.2, we consider a more advanced approach for the transmitter because of its knowledge of the channel matrix.

11.3.1 Channel-Uninformed MIMO Transmitter

If the transmitter does not know anything about the channel, then the best strategy for the transmitter is to use a space–time code that sends spatially uncorrelated signals from the transmit antennas. The transmit covariance matrix is then given by a matrix that is proportional to the identity matrix. We want the total energy transmitted summed over all antennas to be equal to P_{tot}. Thus, we can determine the transmit covariance by determining the unknown constant k by evaluating

$$\begin{aligned} \mathrm{tr}\{\mathbf{P}\} &= P_{tot} \\ &= \mathrm{tr}\{k\,\mathbf{I}\} \\ &= k\,\mathrm{tr}\{\mathbf{I}\} = k \sum_m (\lambda_m\{\mathbf{I}\}) \\ &= k \sum_m 1 = k\,n_t = P_{tot} \\ \mathbf{P} &= \frac{P_{tot}}{n_t}\,\mathbf{I} . \end{aligned} \tag{11.19}$$

If we substitute the form for the transmit covariance $\mathbf{P}$ into Equation (11.17),

$$c = \log_2 \left| \mathbf{I} + \frac{P_{tot}}{n_t} \mathbf{H}\mathbf{H}^H \right| \qquad (11.20)$$

$$= \log_2 \left[\prod_m \lambda_m \left\{ \mathbf{I} + \frac{P_{tot}}{n_t} \mathbf{H}\mathbf{H}^H \right\} \right]$$

$$= \log_2 \left[\prod_m \left(1 + \lambda_m \left\{ \frac{P_{tot}}{n_t} \mathbf{H}\mathbf{H}^H \right\} \right) \right]$$

$$= \sum_m \log_2 \left(1 + \frac{P_{tot}}{n_t} \lambda_m \{\mathbf{H}\mathbf{H}^H\} \right),$$

where $\lambda_m\{\mathbf{M}\}$ is the mth eigenvalue of matrix $\mathbf{M}$. For this formulation, it is clear that the eigenvalues of $\mathbf{H}\mathbf{H}^H$ are important. Related information is contained in the singular values of $\mathbf{H}$. Because it can be difficult to gain access to good estimates of the channel matrix at the transmitter, the bound is often the realistic limit.

11.3.2   Channel-Informed Transmitter

If the transmitter has knowledge of the channel, then it can adjust how the channel is used to maximize the data rate, as indicated in Figure 11.2. This adjustment is done by changing the structure of the spatial transmit covariance matrix. There are multiple ways in which this channel can be provided to the transmitter. An estimate of the channel matrix can be made at the receiver, and this estimate can be communicated back to the transmitter through some communication channel. Another approach is to use channel reciprocity, which exploits the observation that the channel in one direction must be the same as in the other direction up to a time reversal. For a time-division duplex (TDD) system, the channel estimated in one direction can be used to construct an estimate of the channel in the other direction: $\mathbf{H}_{b \to a} = \mathbf{H}_{a \to b}^T$.

By optimizing the transmit covariance matrix $\mathbf{P}$, we maximize the capacity under the constraint that the total power cannot be more than $\text{tr}\{\mathbf{P}\} \leq P_{tot}$. The technical details are a little complicated, but the key is to distribute the power and data

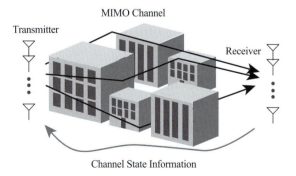

Figure 11.2 Feedback of MIMO channel information.

rate across the modes contained within the channel matrix. We can decompose the channel matrix by using the singular value decomposition (SVD) so that

$$\mathbf{H} = \mathbf{U}\mathbf{S}\mathbf{V}^H \tag{11.21}$$

$$= \sum_m \mathbf{u}_m \alpha_m \mathbf{v}_m^H,$$

where $\mathbf{U}$ and $\mathbf{V}$ are unitary matrices, $\mathbf{S}$ is a diagonal matrix, $\mathbf{u}_m$ and $\mathbf{v}_m$ are the mth columns in their respective unitary matrices, and α_m is the mth singular value located at the mth diagonal element in $\mathbf{S}$. We use α_m here to avoid confusion with the transmitted sequence. A large singular value indicates a strong channel between a particular pair of transmit and receive antenna beamformers. A transmit or receive beamformer is given by a vector of complex amplitudes that when applied to the antenna array, creates a preferred direction for the transmission or reception of the RF signal. In multipath environments, this point may be difficult to interpret. We can rewrite the capacity in Equation (11.17) with the observation (which we state without proof) that the best transmit covariance matrix has the form of

$$\mathbf{P} = \mathbf{V}\operatorname{diag}\{p_1, p_2, \cdots\}\mathbf{V}^H, \tag{11.22}$$

where the matrix $\mathbf{V}$ is the same as that from Equation (11.21) so that

$$c = \log_2 |\mathbf{I} + \mathbf{H}\mathbf{P}\mathbf{H}^H|$$

$$= \log_2 \left| \mathbf{I} + \mathbf{H}\left(\sum_m \mathbf{v}_m p_m \mathbf{v}_m^H\right)\mathbf{H}^H \right|$$

$$= \log_2 \left| \mathbf{I} + \left(\sum_j \mathbf{u}_j \alpha_j \mathbf{v}_j^H\right)\left(\sum_k \mathbf{v}_k p_k \mathbf{v}_k^H\right)\left(\sum_m \mathbf{u}_m \alpha_m \mathbf{v}_m^H\right)^H \right|$$

$$= \log_2 \left| \mathbf{I} + \left(\sum_j \mathbf{u}_j \alpha_j \mathbf{v}_j^H\right)\left(\sum_k \mathbf{v}_k p_k \mathbf{v}_k^H\right)\left(\sum_m \mathbf{v}_m \alpha_m \mathbf{u}_m^H\right) \right|$$

$$= \log_2 \left| \mathbf{I} + \sum_j \mathbf{u}_j \alpha_j p_j \alpha_j \mathbf{u}_j^H \right|$$

$$= \log_2 |\mathbf{I} + \mathbf{U}\operatorname{diag}\{\alpha_1^2 p_1, \alpha_2^2 p_2, \cdots\}\mathbf{U}^H|$$

$$= \log_2 |\mathbf{I} + \operatorname{diag}\{\alpha_1^2 p_1, \alpha_2^2 p_2, \cdots\}|$$

$$= \log_2 \left(\prod_m [1 + \alpha_m^2 p_m] \right)$$

$$= \sum_m \log_2(1 + \alpha_m^2 p_m), \tag{11.23}$$

where we have used the observation that the columns of unitary matrices are orthonormal, $\mathbf{v}_m^H \mathbf{v}_n = \delta_{m,n}$, and that $|\mathbf{I} + \mathbf{A}\mathbf{B}| = |\mathbf{I} + \mathbf{B}\mathbf{A}|$. The next step is to select the power and associated rate that is assigned to each mode. We note that larger values of α_m^2 indicate better modes through the channel. At low signal-to-noise ratio (SNR), it makes sense to put all the power in the mode associated with the best (largest) singular value α_m. As the SNR increases, the compressive nature of the logarithm starts to make putting some of the power into the second mode more attractive. As the power continues to increase, more modes are used. To select how much power to put into each mode, we use the form

$$\lambda_m\{\mathbf{P}\} = p_m = \left(\nu + \frac{1}{\lambda_m\{\mathbf{H}\,\mathbf{H}^H\}} \right)^+ , \tag{11.24}$$

where $\lambda_m\{\mathbf{M}\}$ is the mth eigenvalue of $\mathbf{M}$ and where $(a)^+ = \max(0, a)$. The value of ν is varied until the following condition is satisfied:

$$\sum_m \left(\nu + \frac{1}{\lambda_{+m}\{\mathbf{H}\,\mathbf{H}^H\}} \right)^+ = P_{tot} . \tag{11.25}$$

The value of p_m is then substituted into Equation (11.22).

At the transmitter, we transmit multiple streams of data at different data rates that match the SNR of each matched transmit and receive beamforming pair $\mathbf{u}_m$ and $\mathbf{v}_m$. The transmitted signal has a structure given by

$$\mathbf{s}(t) = \sum_{m=1}^{n_t} \mathbf{v}_m\, s_m(t) , \tag{11.26}$$

where $s_m(t)$ is the complex baseband encoded and modulated data stream associated with the mth beamforming pair. This approach is sometimes called *spatial precoding*. The variance of $s_m(t)$ is set to p_m, and the data rate is bounded by the Shannon limit associated such that

$$r_m \leq \log_2 \left(1 + \frac{\mathbf{u}^H\,\mathbf{H}\,\mathbf{v}\,p_m\,\mathbf{v}^H\mathbf{H}^H\mathbf{v}}{\mathbf{u}^H\,\mathbf{u}} \right)$$
$$= \log_2 \left(1 + p_m\,\|\mathbf{u}^H\,\mathbf{H}\,\mathbf{v}\|^2 \right) , \tag{11.27}$$

where we have used $\mathbf{u}_m^H\,\mathbf{u}_n = \delta_{m,n}$. In general not all beamforming pairs are used because the approach selects the best modes through the channels.

11.4   Space–Time Coding

With multiple transmit antennas, we need an approach to translate a stream of symbols to a sequence transmitted from multiple antennas. These approaches are termed *space–time coding*. Because of the complications involved in accessing the channel state information at the transmitter, it is more common for MIMO systems to operate without it, as is described in Section 11.3.1. Because there is no knowledge of antenna or directional preferences, all antennas are treated the same. There are numerous techniques for implementing space–time codes. There are two extremes in space–time coding approaches, and a continuum between these extremes. One can use the coding approach to maximize data rate, as discussed in Section 11.4.1, or attempt to maximize spatial redundancy, as discussed in Section 11.4.2. The correct choice depends upon the engineering metric of performance that may be weighted more toward data rate or toward reliability.

11.4.1   Maximum-Rate Space–Time Coding

For a given modulation scheme transmitted from each antenna, we can maximize the total number of transmitted symbols by capturing groups of symbols equal to the number of transmit antennas and allocating one symbol to each antenna.[1] The receiver then disentangles symbols sent from each antenna. Typically some forward error correction code would be applied prior to this demultiplexing, as indicated in Figure 11.3. The overall space–time coding rate would then be $r_{st} = r_{fec} n_t$, where r_{st} is the overall coding rate, n_t is the number of transmit antennas, and r_{fec} is the coding rate of the SISO forward error correction code. To be clear, this approach will not provide the maximum throughput for a given transmit power. For example, given a high-rate SISO code, a couple of weak transmit-antenna-to-receiver channels can cause significant performance degradation. For some sequence of the symbols $s_1, s_2, \cdots, s_{n_t}$ at complex baseband (after forward error correction and modulation), the transmit antennas send the vector symbol **s**, which is given by

$$ \mathbf{s} = \begin{pmatrix} s_1 \\ s_2 \\ \vdots \\ s_{n_t} \end{pmatrix}. \tag{11.28} $$

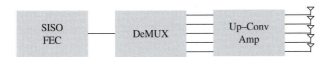

Figure 11.3 Notional five-antenna high-rate MIMO space–time coding approach. The blocks include a typical SISO forward error correction code. This is followed by a demultiplexing block that converts the stream of symbols to five streams, with each antenna transmitting one out of five encoded symbols. The five streams are then up-converted and amplified before being transmitted.

11.4.2   Maximum-Redundancy Space–Time Block Coding

Compared to the high-rate approach, for which the multiple antennas are used to send more symbols through the channel, maximum redundancy space–time block coding guarantees that some version of each symbol is represented at every transmitter. There are a number of these block codes that have various effective coding rates [22, 23]. A typical implementation would employ a forward error correction

---

[1] This approach was termed *V-BLAST* in some of the early literature.

code prior to applying the space–time block code, as indicated in Figure 11.4. The overall space–time coding rate would then be $r_{\text{st}} = r_{\text{fec}}\, r_{\text{block}}$, where r_{st} is the overall coding rate.

Figure 11.4 Notional five-antenna space–time MIMO block coding approach. The blocks include a typical SISO forward error correction code. This is followed by a redundancy block code that represents the same information across all five antennas over a block of symbols. With the block, each symbol is represented on each antenna, potentially multiple times. The five streams are then up-converted and amplified before being transmitted.

As an example, we consider the simple but useful Alamouti space–time block [24]. This coding approach assumes that the transmit system has two antennas. At complex baseband, the code word represents each symbol at each antenna over two successive samples in time. After SISO forward error correction encoding and modulation at complex baseband, two symbols are indicated as s_1 and s_2. We then transmit the following versions of the symbols across the two transmitters over the two samples in time, as indicated by $\mathbf{C}$:

$$\mathbf{C} = \begin{pmatrix} s_1 & -s_2^* \\ s_2 & s_1^* \end{pmatrix}. \tag{11.29}$$

This slightly strange formulation, which employs conjugation and negation, is convenient at the receiver. For the sake of simple discussion, let us consider a 2×1 multiple-input single-output (MISO) flat-fading channel. The channel is represented by the row vector $\underline{\mathbf{h}}$:

$$\underline{\mathbf{h}} = (h_1 \quad h_2). \tag{11.30}$$

For two successive symbols in time, the received signal $\underline{\mathbf{z}}$ at the single receive antenna is then given by the row vector

$$\begin{aligned} \underline{\mathbf{z}} &= \underline{\mathbf{h}}\, \mathbf{C} + \underline{\mathbf{n}} \\ &= (h_1\, s_1 + h_2\, s_2 \quad -h_1\, s_2^* + h_2\, s_1^*) + \underline{\mathbf{n}}, \end{aligned} \tag{11.31}$$

where $\underline{\mathbf{n}}$ is the complex additive noise at the receiver. By assuming that we know the channel at the receiver, we can build the matrix $\mathbf{M}$, which is given by

$$\mathbf{M} = \frac{1}{\|\underline{\mathbf{h}}\|^2} \begin{pmatrix} h_1^* & h_2^* \\ h_2 & -h_1 \end{pmatrix}. \tag{11.32}$$

By using this matrix, we can construct an estimate of the transmitted symbols $\hat{\mathbf{s}}$, which is given by

$$\hat{\mathbf{s}} = (z_1 \quad z_2^*)\,\mathbf{M}$$
$$= [(h_1\,s_1 + h_2\,s_2 \quad (-h_1\,s_2^* + h_2\,s_1^*)^*) + \underline{\mathbf{n}}]\,\mathbf{M}$$
$$= [(h_1\,s_1 + h_2\,s_2 \quad -h_1^*\,s_2 + h_2^*\,s_1)\,\frac{1}{\|\underline{\mathbf{h}}\|^2}\left(\begin{array}{cc} h_1^* & h_2^* \\ h_2 & -h_1 \end{array}\right) + \tilde{\underline{\mathbf{n}}}$$
$$= \frac{1}{\|\underline{\mathbf{h}}\|^2}(h_1 h_1^*\,s_1 + h_2\,h_1^*\,s_2 - h_1^* h_2\,s_2 + h_2^* h_2\,s_1$$
$$h_1 h_2^*\,s_1 + h_2\,h_2^*\,s_2 + h_1^* h_1\,s_2 - h_2^* h_1\,s_1) + \tilde{\underline{\mathbf{n}}}$$
$$= \frac{1}{\|\underline{\mathbf{h}}\|^2}((\|h_1\|^2 + \|h_2\|^2)\,s_1 \quad (\|h_1\|^2 + \|h_2\|^2)\,s_2) + \tilde{\underline{\mathbf{n}}}$$
$$= (s_1 \quad s_2) + \tilde{\underline{\mathbf{n}}}, \tag{11.33}$$

where $\tilde{\underline{\mathbf{n}}} = \underline{\mathbf{n}}\,\mathbf{M}$. If there are multiple receive antennas, the process can be repeated, and the likelihoods of various symbol hypotheses can be combined in the SISO forward error correction decoder.

Problems

11.1 Consider a SIMO channel with an n_{ant}-antenna receiver that observes a complex baseband signal given by

$$\mathbf{z}(t) = \mathbf{h}\,s(t) + \mathbf{n}(t),$$

where t is time, $\mathbf{z}(t) \in \mathbb{C}^{n_{ant} \times 1}$ is the received signal, $\mathbf{h} \in \mathbb{C}^{n_{ant} \times 1}$ is the channel vector of complex attenuation from the transmitter to each receive antenna, $s(t)$ is the transmitted signal with variance P, and $\mathbf{n}(t) \in \mathbb{C}^{n_{ant} \times 1}$ is received noise for which each element is drawn from a zero-mean circularly symmetric unit-variance complex Gaussian distribution. Evaluate the capacity of this system in terms of the spectral efficiency.

11.2 Consider a MIMO channel with n_{tx} transmitting antennas and n_{rx} receiving antennas. The received signal observes a complex baseband signal given by

$$\mathbf{z}(t) = \mathbf{H}\,\mathbf{s}(t) + \mathbf{n}(t),$$

where t is time, $\mathbf{z}(t) \in \mathbb{C}^{n_{rx} \times 1}$ is the received signal, $\mathbf{h} \in \mathbb{C}^{n_{rx} \times n_{tx}}$ is the channel vector of complex attenuation from the transmitter to each receive antenna, $\mathbf{s}(t) \in \mathbb{C}^{n_{tx} \times 1}$ is the transmitted signal, and $\mathbf{n}(t) \in \mathbb{C}^{n_{ant} \times 1}$ is received noise for which each element is drawn from a zero-mean circularly symmetric unit-variance complex Gaussian distribution. If the transmit covariance is given by

$$E\{\mathbf{s}(t)\,\mathbf{s}^H(t)\} = \frac{P}{n_{tx}}\mathbf{I},$$

evaluate the capacity of this system in terms of the spectral efficiency.

11.3 Consider a SIMO channel with an n_{ant}-antenna receiver that observes a complex baseband signal given by

$$\mathbf{z}(t) = \mathbf{h}\,s(t) + \mathbf{n}(t),$$

where t is time, $\mathbf{z}(t) \in \mathbb{C}^{n_{ant} \times 1}$ is the received signal, $\mathbf{h} \in \mathbb{C}^{n_{ant} \times 1}$ is the channel vector of complex attenuation from the transmitter to each receive antenna, $s(t)$ is the transmitted signal, and $\mathbf{n}(t) \in \mathbb{C}^{n_{ant} \times 1}$ is the zero-mean circularly symmetric complex Gaussian noise.

(a) Under the assumptions that the noise at each antenna is independent from other antennas with unit variance and $s(t)$ has unit variance, symbolically evaluate the spatial received covariance matrix $\mathbf{R}$, if it is given by

$$\mathbf{R} = E\{\mathbf{z}(t)\,\mathbf{z}^H(t)\}.$$

(b) Under the above assumptions, and by assuming the channel is normalized so that $\|\mathbf{h}\|^2 = 10$, find the eigenvalues of $\mathbf{R}$.

11.4 Consider a flat-fading channel with a single transmit antenna and four receive antennas. If the channel is given by

$$\mathbf{h} = \begin{pmatrix} 1 \\ 1 \\ 1 \\ 1 \end{pmatrix},$$

what is the optimal spatial receive beamformer that minimizes the mean square error?

11.5 Consider a SIMO system with a single transmitter and four receive antennas. Construct a channel by independently drawing elements from a zero-mean circularly symmetric complex Gaussian distribution. Assume unit variance noise and symbols. The variance of each channel element is the per-antenna receive SNR. Build a simulation that has 100-symbol-long frames of QPSK symbols with unit magnitude. Use the first 20 symbols as training and the later 80 symbols as data. Estimate the adapted spatial beamformer and apply the estimates to 100 randomly generated frames (symbols and channels) at each SNR sample. Repeat the evaluation over a range of per-receive antenna SNR of -10 dB to 10 dB.

(a) By making hard decisions for the QPSK symbols, evaluate the symbol and bit error rate as a function of SNR.

(b) Repeat the analysis under an assumption that an independent single-antenna-transmitter interferer is present with an SNR of 10 dB.

(c) Repeat both previous scenarios but evaluate the signal-to-interference-plus-noise ratio (SINR) at the output of the adapted beamformer. Evaluate the beamformer and apply the beamformer to the signal without noise or interference to estimate the signal variance. Then, apply the same spatial beamformer to the interference-plus-noise signals to estimate the interference-plus-noise variance. The ratio is the SINR. Plot the SINR as a function of the receive SNR.

11.6 The poorly named 90 percent closure outage capacity is evaluated for a block-wise (slow-fading) fluctuating channel such that a given capacity is achieved 90 percent of the time. This metric can be evaluated by drawing random channels for a given average SNR and finding the capacity that is achieved at least 90 percent of the time. For a 4×4 MIMO system, generate channel matrices drawn independently from a unit-variance zero-mean circularly symmetric complex Gaussian distribution. Assume that the units of noise power are defined so that the noise variance is unity. If the transmit covariance matrix is given by $\mathbf{P} = P_0/n_{tx}\,\mathbf{I}$, use Monte Carlo evaluation with simulated channels to evaluate the 90 percent closure outage capacity for

(a) $P_0 = -10$ dB
(b) $P_0 = 0$ dB
(c) $P_0 = 10$ dB

11.7 Repeat Problem 11.6 with the modification of using a rank-1 transmit covariance matrix. Here, the transmit and receive beamformers can be determined by considering the dominant left-hand and right-hand singular vectors of each thrown channel matrix.

11.8 Consider the 4×4 MIMO channel given by

$$\mathbf{H} = \begin{pmatrix} 0.38 - 0.08i & 0.22 + 0.47i & 2.50 + 0.34i & 0.51 + 0.20i \\ 1.29 + 1.05i & -0.92 - 0.85i & 1.95 + 0.73i & -0.04 - 0.55i \\ -1.59 + 0.99i & -0.30 + 0.50i & -0.95 + 0.51i & 0.50 + 0.62i \\ 0.60 + 1.00i & 0.24 + 1.15i & 2.14 - 0.21i & -0.14 - 0.81i \end{pmatrix}.$$

Evaluate the capacity as a function of transmit covariance matrix $\mathbf{P} = P_{tot}/n_{tx}\,\mathbf{I}$, where P_{tot} is the total transmit power under the assumption of unit-variance independent receiver noise.

11.9 Consider the 4×4 MIMO channel given by

$$\mathbf{H} = \begin{pmatrix} 0.38 - 0.08i & 0.22 + 0.47i & 2.50 + 0.34i & 0.51 + 0.20i \\ 1.29 + 1.05i & -0.92 - 0.85i & 1.95 + 0.73i & -0.04 - 0.55i \\ -1.59 + 0.99i & -0.30 + 0.50i & -0.95 + 0.51i & 0.50 + 0.62i \\ 0.60 + 1.00i & 0.24 + 1.15i & 2.14 - 0.21i & -0.14 - 0.81i \end{pmatrix}.$$

Find the optimal spatial water-filling transmit beams and relative powers under the assumption of unit-variance independent receiver noise for the following total transmit powers:

(a) $P_{tot} = -10$ dB
(b) $P_{tot} = 0$ dB
(c) $P_{tot} = 10$ dB

11.10 By using simulation, implement the 2×1 MISO channel version of the Alamouti space–time block code for a stream of QPSK symbols. Evaluate the symbol error rate as a function of SNR (over the range of -5 dB to 15 dB) for a channel for which the elements are independently drawn from a zero-mean circularly symmetric complex Gaussian distribution.

12 Analog Radio Systems

In this chapter we discuss analog radio techniques. Nearly all modern systems employ digital communications. However, for historical reasons, we review these legacy approaches. We introduce linear and angle modulation techniques. Linear modulation approaches include

- double-sideband suppressed-carrier amplitude modulation (DSB-SC AM);
- double-sideband amplitude modulation (DSB AM);
- single-sideband amplitude modulation (SSB AM); and
- vestigial-sideband amplitude modulation (VSB AM).

Angle modulation approaches include

- frequency modulation (FM); and
- phase modulation (PM).

We provide a notional comparison of DSB-SC AM, DSB AM, and PM (and thus FM) in Figure 12.1. To support the analysis of the SSB modulation, we introduce the Hartley modulator, which uses the Hilbert transform. Under the category of angle modulation, we define the PM and FM approaches. We evaluate the spectral content of phase modulated sinusoidal signals. We discuss the use of a phase-lock loop (PLL) to demodulate an FM signal.

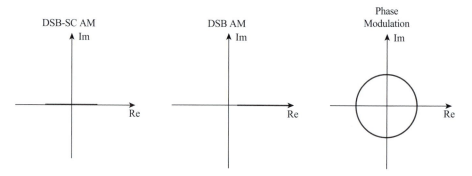

Figure 12.1 Notional representation of the complex baseband support for DSB-SC AM, DSB AM, and PM (and thus FM) approaches.

12.1   Analog Modulation

As we introduce these concepts, we assume the reader is familiar with the digital modulation approaches discussed in Chapter 6. For the digital signals, we typically identified the mapping of bits to complex baseband as *modulation*, and the conversion of complex baseband to passband as *up-conversion*. The nomenclature is often a bit confused in analog communications. For analog communications systems, the modulation and up-conversion are often mixed, and the combination is denoted *modulation*. Because this usage is common in the literature, we employ this nomenclature when convenient, even if it is slightly confusing.

In order to discuss analog modulation, we have to change our perspective to some extent. The paradigm that is based upon the Shannon limit is not particularly relevant. For analog systems, we do not have some bits to send across the channel; rather, we have a continuous analog signal that we wish to transport to the receiver system. Consequently, it is the analog error, which is never driven to zero, that is the measure of performance. While various types of data have been transmitted using this method (for example, the rather complicated color television signal), we focus on the relatively simple transmission that could be applied to monophonic audio. In this case, we have a real analog signal $m(t) \in \mathbb{R}$ that is transmitted. In many legacy systems, the modulated signal includes a strong carrier component. This component is not typically present in digital systems because it is considered an unnecessary waste of power; however, it can be used to simplify the receiver.

12.2   Amplitude Modulation

We consider a few types of amplitude modulation. For some real baseband signal $m(t)$, we can construct the passband signal. A simple and common example is given if we assume that the baseband signal $m(t)$ is an audio signal.

12.2.1   Double-Sideband Suppressed-Carrier Amplitude Modulation

While the name is somewhat cumbersome, double-sideband suppressed-carrier amplitude modulation (DSB-SC AM) is rather familiar because it is similar to that which we normally do for digital communications. It is referred to as *suppressed-carrier* because a carrier contribution is not added to the modulated signal, which is amusing to modern sensibilities because that would be the expected waveform design. Similar to Equation (6.9), we have

$$s_{pb}(t) = \mathfrak{R}\{s(t)\, e^{2\pi f_c t}\} = \mathfrak{R}\{s(t)\} \cos(2\pi f_c t) - \mathfrak{I}\{s(t)\} \sin(2\pi f_c t), \qquad (12.1)$$

where we have invoked Equation (5.3). If we assume that $s(t) = m(t)$ and that $m(t) \in \mathbb{R}$, then

$$s_{pb}(t) = m(t)\, \cos(2\pi f_c t). \qquad (12.2)$$

In Figure 12.2, we display an example of DSB-SC AM. It may be worth drawing a comparison to the binary-phase-shift keying (BPSK) modulation that we depict in Figure 6.3. In some sense, BPSK is the digital sibling of DSB-SC AM. We notice that the baseband signal causes the sign of the carrier signal to flip. Consequently, the demodulation must be sufficiently sophisticated to track this phase inversion. We generally assume that this is not a challenge for digital systems. However, for the legacy systems that used analog modulations, such coherent receivers were often considered to be too complicated.

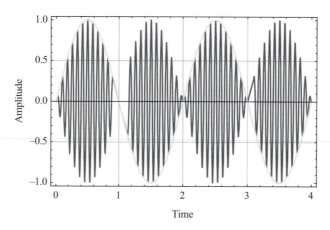

Figure 12.2 Notional example of double-sideband suppressed-carrier amplitude modulation. For this example, the baseband signal is a sine wave.

If we indicate the baseband spectrum of $m(t)$ as $M(f)$, the spectrum of the up-converted modulated signal is given by

$$
\begin{aligned}
\mathcal{F}\{s_{pb}(t)\} &= \mathcal{F}\{m(t) \cos(2\pi f_c t)\} \\
&= M(f) * \frac{[\delta(f - f_c) + \delta(f + f_c)]}{2} \\
&= \int_{-\infty}^{\infty} dv\, M(f - v) \frac{[\delta(v - f_c) + \delta(v + f_c)]}{2} \\
&= \frac{[M(f - f_c) + M(f + f_c)]}{2}.
\end{aligned}
\tag{12.3}
$$

12.2.2   Double-Sideband Amplitude Modulation

As a modification of DSB-SC AM, a tone is often added to the spectrum at the carrier frequency, which is the equivalent of adding a DC offset at baseband, which we call double-sideband amplitude modulation (DSB AM). In Figure 12.3, we see the spectral difference between DSB AM and DSB-SC AM. This is analog modulation's analog-to-digital training sequences. The component helps the receiver lock onto the signal. The passband signal is then given by

$$
s_{pb}(t) = [A + m(t)] \cos(2\pi f_c t),
\tag{12.4}
$$

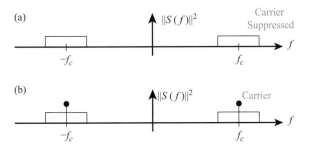

Figure 12.3 Power spectral density of the up-converted notional spectrum $M(f)$ for both (a) double-sideband suppressed-carrier amplitude modulation and (b) double-sideband amplitude modulation.

where A is the tone added at the carrier frequency. In Figure 12.4, we depict the double-sideband amplitude modulation that introduces a significant DC offset in the baseband signal. This approach introduces a power inefficiency; however, the addition of the carrier signal to the waveform allows the receiver approach to be extremely simple. We observe a shifted version of the baseband signal in the envelope of the passband signal. A simple receiver can be built by applying a low-pass filter to a rectified signal, as we show in Figure 12.5. While more sophisticated versions of this circuit are often employed, this simple circuit does work with the right choice of resistor and capacitor.

12.2.3 Single-Sideband Amplitude Modulation

To increase spectral efficiency, we can remove one side of the baseband spectrum, which we call single-sideband amplitude modulation (SSB AM). If we assume

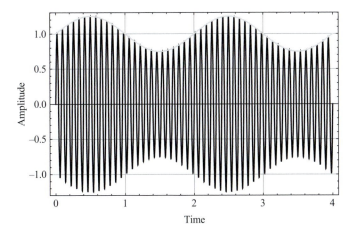

Figure 12.4 Notional example of double-sideband amplitude modulation (DSB AM) in which the carrier is added. For this example, the baseband signal is a sine wave.

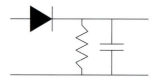

Figure 12.5 Notional demodulation circuit for DSB AM.

that our baseband signal $m(t)$ is real, we observe a conjugate frequency reflection symmetry about DC at baseband. We can use Equation (15.14) to show that $M^*(f) = M(-f)$:

$$M(f) = \mathcal{F}\{m(t)\} \tag{12.5}$$

$$M^*(f) = \left(\int_{-\infty}^{\infty} dt\, e^{-i\,2\pi f t}\, m(t) \right)^*$$

$$= \int_{-\infty}^{\infty} dt\, e^{i\,2\pi f t}\, m(t)$$

$$= \int_{-\infty}^{\infty} dt\, e^{-i\,2\pi\,(-f) t}\, m(t)$$

$$= M(-f). \tag{12.6}$$

Consequently, for a real baseband signal $m(t)$, the information on one side of the spectrum is redundant with the information on the other side.

Because of this spectral conjugate frequency reflection symmetry about the carrier frequency, for the sake of spectral efficiency, we can filter the signal on one side of that spectrum. At passband, this filter can be inflexible and challenging to build. Building the filter becomes easier if we use a superheterodyne modulator. The filter can be implemented at intermediate frequency (IF).

A formal approach to constructing a filter is to use a Hartley modulator that employs a Hilbert transform[1] of $m(t) \to \tilde{m}(t)$. We can use this transform to select only half of the spectrum. We observe that if we only want the positive portion of the baseband signal $M_+(f)$, then we have

$$M_+(f) = \begin{cases} M(f) & f > 0 \\ 0 & \text{otherwise.} \end{cases}$$

$$= \frac{1}{2}[1 + \text{sgn}(f)]\, M(f) \tag{12.7}$$

$$m_+(t) = \mathcal{F}^{-1}\left\{ \frac{M(f)}{2} + \frac{1}{2}\text{sgn}(f)\, M(f) \right\}$$

$$= \frac{m(t)}{2} + \frac{1}{2}\mathcal{F}^{-1}\{\text{sgn}(f)\} * m(t), \tag{12.8}$$

---

[1] For those who are interested in transform theory, there is an interesting formal similarity between the Hilbert transform and the Stieltjes transform.

where the sign function and its inverse Fourier transform are depicted in Figure 12.6. The inverse Fourier transform of the sign function is given by

$$\mathcal{F}^{-1}\{\text{sgn}(f)\} = i\,\frac{1}{\pi\,t}. \tag{12.9}$$

We define the convolution of some function, in this case the baseband signal $m(t)$, with the $1/(\pi\,t)$ to be the Hilbert transform of $m(t)$, which we denote $\tilde{m}(t)$. The Hilbert transform is given by

$$\tilde{m}(t) = \frac{1}{\pi\,t} * m(t)$$

$$= \frac{1}{\pi} \int_{-\infty}^{\infty} d\tau \, \frac{1}{\tau - t}\, m(\tau) \tag{12.10}$$

$$m_+(t) = \frac{1}{2}m(t) + i\,\frac{1}{2}\tilde{m}(t). \tag{12.11}$$

We note that $\tilde{m}(t)$ is real, so the i implies the use of the $\sin(2\pi f_c\,t)$ term in the up-conversion given in Equation (5.4). Consequently, the up-converted signal at passband is given by

$$s_{ssb}(t) = \frac{m(t)}{2}\cos(2\pi f_c\,t) - \frac{\tilde{m}(t)}{2}\sin(2\pi f_c\,t). \tag{12.12}$$

This equation has formal similarities to the quadrature modulation that we discussed in Section 5.2. While more spectrally efficient, this signal is more difficult to demodulate than double-sideband signals.

Because the Hilbert transform is a convolution, it is effectively a filter. Somewhat unfortunately, the "function" is not well-behaved as it has terms that approach infinity. However, we can approximate the function with sufficient accuracy to be useful. Because we do our processing digitally, we can implement these approximate filters digitally. Specifically, we can use Parks–McClellan finite-impulse-response (FIR) filter optimization to make such a filter. As an example, a 10-tap FIR filter is given by

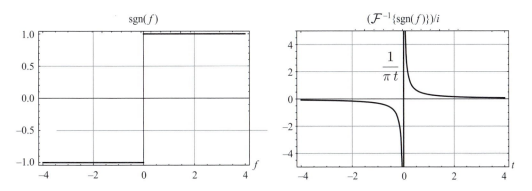

Figure 12.6 Sign function and its inverse Fourier transform.

$$\underline{h} \approx \begin{bmatrix} -0.108 \\ -0.000 \\ -0.183 \\ 0.000 \\ -0.626 \\ 0 \\ 0.626 \\ 0.000 \\ 0.183 \\ 0.000 \\ 0.108 \end{bmatrix}^T . \tag{12.13}$$

Consider an example of 1000 random samples drawn from a zero-mean real Gaussian distribution. By convolving this structure with a random real sequence and taking some care to align the filtered sequence with the original sequence, we produce a baseband signal that is approximately single-sided. In Figure 12.7, we depict the energy spectral density of both the original sequence and the phase-shifted filtered sequence.

12.2.4   Vestigial-Sideband Amplitude Modulation

The vestigial-sideband amplitude modulation (VSB AM) is essentially a sloppy version of SSB AM. It is of particular value if the underlying signal being modulated has significant energy near DC. Because it is difficult to implement a sharp filter, VSB AM trims one side of the baseband signal without getting spectrally close to DC.

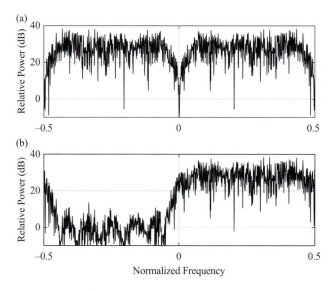

Figure 12.7 Estimated energy spectral density of a random Gaussian sequence (a) and sum of phase-shifted Hilbert transformed sequence and the original sequence (b).

12.3   Phase and Frequency Modulation

We consider analog PM and FM, which are commonly used in legacy radio systems. PM and FM are specific forms of the more general idea of angle modulation. They have a strong connection between digital phase modulation (for example, 8-PSK, discussed in Section 6.1.2). The passband signal is given by

$$s_{pb} = \Re\{e^{i\,2\pi f_c t}\, e^{i\phi(t)}\} = \Re\{e^{i\,[2\pi f_c t + \phi(t)]}\}$$
$$= \cos[2\pi f_c t + \phi(t)] = \cos[\theta(t)], \qquad (12.14)$$

where $\theta(t) = 2\pi f_c t + \phi(t)$. Theoretically, a frequency exists for all time, so the notion of a time-varying frequency is slightly suspect. Nonetheless, it is convenient sometimes to consider the idea of the instantaneous frequency, which is proportional to the derivative of phase with respect to time. If the derivative were constant, then we would have signal at a given frequency. This idea of instantaneous frequency is only a convenient fiction, and it can lead to errors if taken too seriously. The instantaneous frequency $f_{inst}(t)$ is given by

$$f_{inst}(t) = \frac{1}{2\pi}\frac{\partial}{\partial t}\theta(t). \qquad (12.15)$$

For PM, the instantaneous frequency is given by

$$f_{inst}(t) = \frac{1}{2\pi}\frac{\partial}{\partial t}(2\pi f_c t + \phi(t))$$
$$= f_c + \frac{1}{2\pi}\frac{\partial}{\partial t}\phi(t). \qquad (12.16)$$

Phase and frequency modulation are related. In both cases, the signal is essentially a phase modulation. For direct PM, the phase is given by

$$\phi(t) = k_p\, m(t), \qquad (12.17)$$

where k_p is a constant that converts the signal amplitude to the modulated phase. For frequency modulation, the phase is given by

$$\phi(t) = 2\pi k_f \int_{t_0}^{t} d\tau\, m(t), \qquad (12.18)$$

where k_f is a constant that converts the integrated signal amplitude to the modulated phase and t_0 is some initial time.

The modulation index is a parameter that is sometimes used to characterize the modulation. This parameter indicates the maximum deviation in phase. For PM, the modulation index is given by

$$\beta_p = k_p\, \max_t\{\|m(t)\|\}. \qquad (12.19)$$

For FM, it is given by

$$\beta_f = \frac{k_f\, \max_t\{\|m(t)\|\}}{F}, \qquad (12.20)$$

where F is the maximum frequency content of the $m(t)$.

12.4   Sinusoidal Phase Modulation

In general, it is difficult to construct a closed-form evaluation of the spectrum of a phase modulation. A sinusoidal phase modulation is an interesting example for which we can calculate the spectrum. We assume a passband signal, which is given by

$$s_{pb}(t) = \cos[2\pi f_c\, t + \beta\, \sin(2\pi f_m\, t)]$$
$$= \Re\{e^{i\, 2\pi f_c\, t}\, e^{i\beta\, \sin(2\pi f_m\, t)}\}\,, \tag{12.21}$$

where f_m is the frequency of the baseband modulation. We will focus on the complex baseband representation $s(t)$, which is given by

$$s(t) = e^{i\beta\, \sin(2\pi f_m\, t)}\,. \tag{12.22}$$

Because this is a periodic signal, we can use a Fourier series to represent the spectrum of this baseband signal $s(t)$. From Section 15.5, the coefficients are given by

$$c_k = \frac{1}{T}\int_{-T/2}^{T/2} dt\, e^{i\, 2\pi\, k\, t/T}\, s(t)\,, \tag{12.23}$$

where the duration of periodicity for this example is given by $T = 1/f_m$. The Fourier series coefficients can be explicitly evaluated in terms of Bessel functions and are given by

$$c_k = f_m \int_{-T/2}^{T/2} dt\, e^{i\, 2\pi\, k\, t f_m}\, e^{i\beta\, \sin(2\pi f_m\, t)}$$
$$= \frac{1}{2\pi}\int_{0}^{2\pi} dx\, e^{-ixk}\, e^{i\beta\, \sin(x)}\,; \quad x = 2\pi f_m\, t$$
$$= \frac{1}{2\pi}\int_{0}^{2\pi} dx\, e^{i\beta\, \sin(x) - ixk} = J_k(\beta)\,, \tag{12.24}$$

where $J_m(\beta)$ is a Bessel function (defined in Section 13.10.3). Thus, the function $e^{i\beta\, \sin(2\pi f_m\, t)}$ is given by

$$e^{i\beta\, \sin(2\pi f_m\, t)} = \sum_{k=-\infty}^{\infty} J_k(\beta)\, e^{i\, 2\pi f_m\, t k}\,. \tag{12.25}$$

The passband signal is then given by

$$s_{pb}(t) = \Re\{e^{i\, 2\pi f_c\, t}\, e^{i\beta\, \sin(2\pi f_m\, t)}\}$$
$$= \Re\left\{e^{i\, 2\pi f_c\, t} \sum_{k=-\infty}^{\infty} J_k(\beta)\, e^{i\, 2\pi f_m\, t k}\right\}$$
$$= \sum_{k=-\infty}^{\infty} J_k(\beta)\, \cos[2\pi\, t\, (f_m\, k + f_c)]\,. \tag{12.26}$$

As an example, we plot a portion of the spectral amplitudes in Figure 12.8.

12.5   Analog FM Demodulation

We consider a PLL technique to demodulate a frequency-modulated signal [25]. This approach is related to the one used by the frequency synthesizer discussed in Section 5.8. We depict this PLL circuit in Figure 12.9. We can extract the baseband modulation signal from the error signal voltage that drives the voltage-controlled oscillator (VCO). A PM demodulator can be constructed by integrating the FM demodulation output with respect to time. We show that, when locked, the deviation from the carrier frequency or (IF) at the input of the PLL is approximately the deviation from the carrier frequency (or IF) at the VCO output.

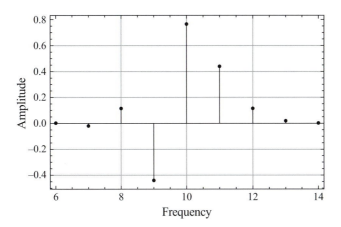

Figure 12.8 Spectral amplitudes of sinusoidal phase modulation with a carrier frequency of 10, a modulation frequency of 1, and a sinusoidal amplitude of $\beta = 1$. The heights indicate the area under the impulses.

This relationship is explicitly expressed as $k_f\, m(t) \approx k_v\, v(t)$, where k_f relates the baseband modulation signal voltage to frequency and k_v relates the VCO input voltage $v(t)$ to frequency.

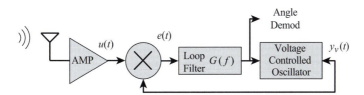

Figure 12.9 Notional PLL demodulator of an FM signal.

We depict a simplified version of the demodulation circuit in Figure 12.9. It is common to use a superheterodyne receiver so that the input is actually an IF that employs a band-pass filter to remove unwanted signals. For simplicity of discussion, we consider a reduced complexity version. First, the received signal is amplified. We denote the output of this amplifier by $u(t)$. This signal is mixed with the output of the

VCO, which we denote as $y_v(t)$. The output of the mixer contains signals around the frequencies of the sum and difference of frequencies in $u(t)$ and $y_v(t)$. If the PLL is locked, then there are frequency components near DC and twice the carrier frequency (or IF) at the output of the mixer. The doubled frequency component is not useful, so we remove it with a low-pass filter characterized by $G(f)$.

For our frequency modulation $m(t)$, we can produce the corresponding phase modulation $\phi(t)$ and time-domain signal, which are given by

$$\phi(t) = 2\pi\, k_f \int_0^t d\tau\, m(\tau) \tag{12.27}$$

$$u(t) = A\, \cos[2\pi\, f_c\, t + \phi(t)], \tag{12.28}$$

where k_f is the constant that maps input signal voltage to phase.

If we assume that the PLL is working properly, then the carrier frequency (or IF) f_c will be locked to the input frequency, and the output of the VCO $f_v(t)$ will be near that carrier frequency. Thus, we can express the PLL output frequency $f_v(t)$ as

$$f_v(t) = f_c + k_v\, v(t) \tag{12.29}$$

$$y_v(t) = A_v\, \sin[2\pi\, f_c\, t + \phi_v(t)] \tag{12.30}$$

$$\phi_v(t) = 2\pi\, k_v \int_0^t d\tau\, v(t), \tag{12.31}$$

where $v(t)$ is the input voltage for the VCO from the carrier frequency, k_v is the voltage to frequency conversion constant for the VCO, and $\phi_v(t)$ is the phase at the output of the VCO. If the low-pass filter $G(f)$ removes the doubled frequencies and any other harmonics and passes the component near DC without significant distortion, then we can ignore the higher-frequency components at the output of an ideal mixer. The output of the mixer is given by

$$\begin{aligned} e(t) &= u(t)\, y_v(t) \\ &= A\, \cos[2\pi\, f_c\, t + \phi(t)]\, A_v\, \sin[2\pi\, f_c\, t + \phi_v(t)] \\ &= \frac{1}{2} A A_v\, \sin[\phi(t) - \phi_v(t)] + h.f.c., \end{aligned} \tag{12.32}$$

where $h.f.c.$ represents the high-frequency components that will be removed by the filter. We assume that the PLL is working well, so the phase error – that is, the phase of the low-frequency component of $e(t)$, which we denoted $\phi_e(t)$ – is small so that the sine of the phase error is essentially the phase error:

$$\phi_e(t) \approx \phi(t) - \phi_v(t). \tag{12.33}$$

This linearization assumption is not always valid, but we assume it to be true throughout the rest of this discussion. We observe that we can produce a phase version of Figure 12.9 and focus on the phase at each point in the PLL. We depict this phase version in Figure 12.10.

The phase error is given by

$$\begin{aligned} \phi_e(t) &= \phi(t) - \phi_v(t) \\ &= \phi(t) - 2\pi\, k_v \int_0^t d\tau\, v(t), \end{aligned} \tag{12.34}$$

where the phase of the output of the VCO is produced by integrating the instantaneous frequency since the start of time. We now consider the time derivative of both sides to remove the integral. Furthermore, we can now introduce the effect of the filter into the PLL relationship. Our phase difference equation is then given by

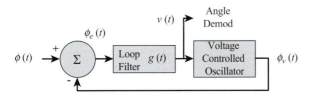

Figure 12.10 Notional PLL demodulator of an FM signal from a phase perspective.

$$\frac{\partial}{\partial t}\phi_e(t) = \frac{\partial}{\partial t}\phi - 2\pi\,k_v\,v(t) \tag{12.35}$$

$$= \frac{\partial}{\partial t}\phi - 2\pi\,k_v\int_0^t d\tau\,\phi_e(\tau)\,g(t-\tau). \tag{12.36}$$

By considering the Fourier transform of both sides of the equation, we can simplify the convolution and solve for the phase error directly. This form is given by

$$\mathcal{F}\left\{\frac{\partial}{\partial t}\phi - 2\pi\,k_v\int_0^t d\tau\,\phi_e(\tau)\,g(t-\tau)\right\}$$
$$= i\,2\pi\,f\,\Phi_e(f) = i\,2\pi\,f\,\Phi(f) - i\,2\pi\,k_v\,\Phi_e(f)\,G(f) \tag{12.37}$$

$$\Phi_e(f) = \frac{1}{1 + \frac{k_v}{if}G(f)}\,\Phi(f), \tag{12.38}$$

where $\Phi_e(f)$ and $\Phi(f)$ are the Fourier transforms of the phase error and input phase, respectively. By noting that the Fourier transform of $v(t)$, which we denote $V(t)$, is given by applying the effect of the filter $G(f)$,

$$V(f) = \Phi_e(f)\,G(f)$$
$$= \frac{G(f)}{1 + \frac{k_v}{if}G(f)}\,\Phi(f). \tag{12.39}$$

We consider the high gain limit $\|G(f)\,k_v/f\| \gg 1$ so that $V(f)$ and its inverse Fourier transform $v(t)$ are given by

$$V(f) = \frac{1}{\left(\frac{k_v}{if}\right)}\,\Phi(f)$$
$$= \frac{2\pi}{2\pi}\frac{if}{k_v}\,\Phi(f) \tag{12.40}$$

$$v(t) = \frac{1}{2\pi\,k_v}\frac{\partial}{\partial t}\phi(t). \tag{12.41}$$

By recalling that $\phi(t)$ is given by

$$\phi(t) = 2\pi \, k_f \int_0^t d\tau \, m(\tau), \tag{12.42}$$

we can relate

$$v(t) \approx \frac{1}{2\pi \, k_v} \frac{\partial}{\partial t} \phi(t) \tag{12.43}$$

$$= \frac{k_f}{k_v} m(t)$$

$$\hat{m}(t) = \frac{k_v}{k_f} v(t), \tag{12.44}$$

where $\hat{m}(t)$ is our estimate of $m(t)$.

Problems

12.1 For a real modulated signal $m(t)$, show that the baseband spectrum is conjugate symmetric about DC, $M^*(f) = M(-f)$.

12.2 In simulation, construct a DSB AM signal from a baseband signal given by $3 + \sin(2\pi t)$. By using a modulation frequency that is 100 times higher, construct the passband signal. Oversample the signal at a factor of 10. Construct a rectification and filtering algorithm that recovers the sinusoidal component of the baseband signal.

12.3 Consider the analog modulation of real message functions $m_1(t)$ and $m_2(t)$ modulated to a carrier frequency f_c. Assume that the Fourier transform of the modulated signals at passband are $U_1(f)$ for $m_1(t)$ and $U_2(f)$ for $m_2(t)$.
(a) Under amplitude modulation of $m_s(t) = m_1(t) + m_2(t)$ (DSB-SC, for example), what is the Fourier transform of the passband signal? Explain.
(b) Under phase modulation and with this information, can you explicitly express the Fourier transform of the passband signal of $m_s(t) = m_1(t) + m_2(t)$? Explain.

12.4 Consider the frequency modulation of real message functions $m(t)$ up-converted to a carrier frequency f_c. Assume that $m(t)$ can only take on the values of 1 or -1 with equal likelihood, that the deviation factor is given by $k_f = f_m$ such that $f_m \ll f_c$, and that changes in $m(t)$ values occur very infrequently.
(a) Plot approximately the passband power spectral density.
(b) How frequently could $m(t)$ change such that it would still be easy to tell the difference? Explain.

12.5 Consider the sinusoidal phase modulation. Evaluate the spectral amplitudes for this modulation if the carrier frequency is 10 and for the following modulation frequencies: for β of
(a) $\beta = 0.2$
(b) $\beta = 0.5$
(c) $\beta = 1$
(d) $\beta = 1.5$

Part II

Mathematical Background

13　Useful Mathematics

Yes, mathematics may be difficult on occasion, but doing anything technically interesting without math is impossible. Learning mathematics is akin to learning a language, so you should expect that it will take significant practice to become accomplished.

In this chapter we discuss a number of useful definitions and relationships used throughout the text by expanding upon our discussions in Chapter 1. Some of these definitions are drawn directly from Reference [21], which has a more thorough discussion of these topics. In the remainder of the text, it is assumed that the reader has familiarity with these topics. In general, the relationships are stated without proof, and the reader is directed to dedicated mathematical texts for further detail [4, 26–30]. We provide a quick review of linear algebra, including tools for matrix inversion. We discuss special matrices such as Hermitian, unitary, and Toeplitz. We review the concepts of norm, trace, and determinant. We introduce special products such as the Hadamard and Kronecker. We discuss matrix decomposition techniques. We introduce techniques for optimizing real functions with respect to complex parameters. We review multivariate calculus. We review a number of useful special functions: sinc, gamma, Bessel, error, Gaussian Q, and Marcum Q.

13.1　Vectors and Matrices

A vector space is an important concept that we employ throughout the text. A vector has some direction and length in our space. It is relatively easy for us to imagine a vector in a real three-dimensional space that corresponds to the physical space in which we live, as we depict in Figure 13.1. Many of the problems in communications extend this idea to higher-dimensional spaces.

We will assume that vector spaces employed within the text satisfy the requirements of a (usually Cartesian) Hilbert space, that is, vectors have meaningful inner products and norms. We indicate a vector by a bold lowercase letter. For example, a column n-vector of complex values is indicated by

$$\mathbf{a} \in \mathbb{C}^{n \times 1}. \tag{13.1}$$

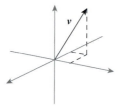

Figure 13.1 Notional depiction of a real 3-vector v.

The $n \times 1$ contained within the superscript of $\mathbb{C}^{n \times 1}$ indicates that the object has n elements along a row and is one column wide. As an example, we could have a 3×1 vector given by

$$\mathbf{a} = \begin{pmatrix} 2 \\ 3 - i \\ 2i \end{pmatrix}. \tag{13.2}$$

A row n-vector is indicated by a bold lowercase letter with an underscore, such as

$$\underline{\mathbf{a}} \in \mathbb{C}^{1 \times n}, \tag{13.3}$$

which has all n of its elements along one row. As an example, we could have a 1×3 row vector given by

$$\underline{\mathbf{a}} = \begin{pmatrix} 2 & 3 - i & 2i \end{pmatrix}. \tag{13.4}$$

On occasion, we need to extract one element of a vector. The mth element in $\mathbf{a}$ is denoted

$$(\mathbf{a})_m \text{ or } \{\mathbf{a}\}_m. \tag{13.5}$$

We indicate a matrix with m rows and n columns by a bold uppercase letter, for example,

$$\mathbf{M} \in \mathbb{C}^{m \times n}$$

$$= \begin{pmatrix} M_{1,1} & M_{1,2} & M_{1,3} & \cdots & M_{1,n} \\ M_{2,1} & M_{2,2} & M_{2,3} & \cdots & M_{2,n} \\ M_{3,1} & M_{3,2} & M_{3,3} & & \\ \vdots & & & \ddots & \\ M_{m,1} & M_{m,2} & M_{m,3} & \cdots & M_{m,n} \end{pmatrix}, \tag{13.6}$$

where $M_{p,q}$ is the element at the pth row and qth column of $\mathbf{M}$. If there is potential ambiguity, we find it useful to use the notation that operates on the matrix for the pth row and qth column of $\mathbf{M}$ that is given by

$$(\mathbf{M})_{p,q} \text{ or } \{\mathbf{M}\}_{p,q}. \tag{13.7}$$

We indicate the complex conjugate of vectors and matrices by

$$
\mathbf{a}^* = \begin{pmatrix} a_1 \\ a_2 \\ \vdots \\ a_m \end{pmatrix}^* = \begin{pmatrix} a_1^* \\ a_2^* \\ \vdots \\ a_m^* \end{pmatrix}
\tag{13.8}
$$

and

$$
\mathbf{M}^* = \begin{pmatrix}
M_{1,1}^* & M_{1,2}^* & M_{1,3}^* & \cdots & M_{1,n}^* \\
M_{2,1}^* & M_{2,2}^* & M_{2,3}^* & \cdots & M_{2,n}^* \\
M_{3,1}^* & M_{3,2}^* & M_{3,3}^* & & \\
\vdots & & & \ddots & \\
M_{m,1}^* & M_{m,2}^* & M_{m,3}^* & \cdots & M_{m,n}^*
\end{pmatrix},
\tag{13.9}
$$

where conjugation operates on each element independently. We indicate the transpose of vectors and matrices by

$$
\mathbf{a}^T = \begin{pmatrix} a_1 \\ a_2 \\ \vdots \\ a_m \end{pmatrix}^T = (a_1 \quad a_2 \quad \cdots \quad a_m)
\tag{13.10}
$$

and

$$
\mathbf{M} \in \mathbb{C}^{m \times n} \text{ and } \mathbf{M}^T \in \mathbb{C}^{n \times m}
$$

$$
\mathbf{M}^T = \begin{pmatrix}
M_{1,1} & M_{1,2} & M_{1,3} & \cdots & M_{1,n} \\
M_{2,1} & M_{2,2} & M_{2,3} & \cdots & M_{2,n} \\
M_{3,1} & M_{3,2} & M_{3,3} & & \\
\vdots & & & \ddots & \\
M_{m,1} & M_{m,2} & M_{m,3} & \cdots & M_{m,n}
\end{pmatrix}^T
$$

$$
= \begin{pmatrix}
M_{1,1} & M_{2,1} & M_{3,1} & \cdots & M_{m,1} \\
M_{1,2} & M_{2,2} & M_{3,2} & \cdots & M_{m,2} \\
M_{1,3} & M_{2,3} & M_{3,3} & & \\
\vdots & & & \ddots & \\
M_{1,n} & M_{2,n} & M_{3,n} & \cdots & M_{m,n}
\end{pmatrix},
\tag{13.11}
$$

respectively. We indicate the Hermitian conjugate (or equivalently conjugate transpose) of vectors and matrices by

$$
\mathbf{a}^H = (\mathbf{a}^T)^* = (\mathbf{a}^*)^T \quad \text{and}
$$
$$
\mathbf{M}^H = (\mathbf{M}^T)^* = (\mathbf{M}^*)^T,
\tag{13.12}
$$

respectively.

It is sometimes useful to partition a vector or matrix into pieces. For example, we partition the matrix $\mathbf{M}$ into

$$\mathbf{M} = \begin{pmatrix} \mathbf{A} & \mathbf{b} \\ \mathbf{c}^H & d \end{pmatrix}, \tag{13.13}$$

where $\mathbf{b}$ and $\mathbf{c}$ are of the appropriate size to match the number of rows and columns, respectively, and d is a scalar. As a concrete example, consider the scenario in which the $\mathbf{M}$ is a 3×3 matrix. In this scenario, the matrix can be specified as

$$\mathbf{M} = \begin{pmatrix} m_{1,1} & m_{1,2} & m_{1,3} \\ m_{2,1} & m_{2,2} & m_{2,3} \\ m_{3,1} & m_{3,2} & m_{3,3} \end{pmatrix} \tag{13.14}$$

so that, for the above example,

$$\mathbf{A} = \begin{pmatrix} m_{1,1} & m_{1,2} \\ m_{2,1} & m_{2,2} \end{pmatrix} \tag{13.15}$$

$$\mathbf{b} = \begin{pmatrix} m_{1,3} \\ m_{2,3} \end{pmatrix} \tag{13.16}$$

$$\mathbf{c} = \begin{pmatrix} m_{3,1}^* \\ m_{3,2}^* \end{pmatrix} \tag{13.17}$$

$$d = m_{3,3}. \tag{13.18}$$

The Hermitian conjugate of this partitioned matrix is then given by

$$\mathbf{M}^H = \begin{pmatrix} \mathbf{A}^H & \mathbf{c} \\ \mathbf{b}^H & d^* \end{pmatrix}. \tag{13.19}$$

We indicate a diagonal matrix by

$$\mathrm{diag}\{a_1, a_2, a_3, \cdots, a_n\} = \begin{pmatrix} a_1 & 0 & 0 & \cdots & 0 \\ 0 & a_2 & 0 & & \\ 0 & 0 & a_3 & & \\ \vdots & & & \ddots & \\ 0 & & & & a_n \end{pmatrix}. \tag{13.20}$$

The transpose of a square diagonal matrix is itself.

The vector operation is a clumsy concept that maps matrices to vectors. It could be avoided by employing tensor operations. However, it is sometimes convenient for the sake of implementation to consider explicit conversions between matrices and vectors. The vector operation extracts elements along each column before moving to the next column. We denote the vector operation of matrix $\mathbf{M} \in \mathbb{C}^{M \times N}$ by $\mathrm{vec}(\mathbf{M})$, and we define it to be

$$\{\mathrm{vec}(\mathbf{M})\}_{(n-1)M+m} = \{\mathbf{M}\}_{m,n}. \tag{13.21}$$

As a concrete example, if the 2×2 matrix $\mathbf{M}$ is specified by

$$\mathbf{M} = \begin{pmatrix} m_{1,1} & m_{1,2} \\ m_{2,1} & m_{2,2} \end{pmatrix}, \tag{13.22}$$

then the vec($\mathbf{M}$) reads down the columns of $\mathbf{M}$ and is given by

$$\text{vec}(\mathbf{M}) = \begin{pmatrix} m_{1,1} \\ m_{2,1} \\ m_{1,2} \\ m_{2,2} \end{pmatrix}. \tag{13.23}$$

13.2 Vector and Matrix Products

We indicate the dot product, which is the inner product between real N-dimensional column vectors, by

$$\mathbf{a} \cdot \mathbf{b} = \mathbf{a}^T \mathbf{b} = \sum_{m=1}^{N} \{\mathbf{a}\}_m \{\mathbf{b}\}_m. \tag{13.24}$$

As an explicit example, the inner product of two real column vectors is given by

$$\begin{pmatrix} 2 \\ 3 \\ -2 \end{pmatrix}^T \begin{pmatrix} 1 \\ -1 \\ -2 \end{pmatrix} = 2 - 3 + 4 = 3. \tag{13.25}$$

We denote the inner product for complex N-dimensional column vectors (or, more precisely, the Hermitian inner product) by

$$\mathbf{a}^H \mathbf{b} = \sum_{m=1}^{N} \{\mathbf{a}\}_m^* \{\mathbf{b}\}_m. \tag{13.26}$$

As an explicit example, the inner product of two complex column vectors is given by

$$\begin{pmatrix} i \\ 1 \\ i \end{pmatrix}^H \begin{pmatrix} 1 \\ 1+i \\ i \end{pmatrix} = -i + (1+i) - i(i) = 2. \tag{13.27}$$

We indicate the outer product of two vectors $\mathbf{a} \in \mathbb{C}^{M \times 1}$ and $\mathbf{b} \in \mathbb{C}^{N \times 1}$ by

$$\mathbf{a}\,\mathbf{b}^H. \tag{13.28}$$

This outer product produces a matrix of the size $M \times N$. The mth, nth element of the matrix is given by

$$\{\mathbf{a}\,\mathbf{b}^H\}_{m,n} = \{\mathbf{a}\}_m \{\mathbf{b}\}_n^*. \tag{13.29}$$

As an explicit example, the outer product of two complex column vectors is given by

$$\begin{pmatrix} i \\ 1 \\ i \end{pmatrix} \begin{pmatrix} 1 \\ 1+i \\ i \end{pmatrix}^{H} = \begin{pmatrix} i \\ 1 \\ i \end{pmatrix} \begin{pmatrix} 1 & 1-i & -i \end{pmatrix}$$

$$= \begin{pmatrix} i & i+1 & 1 \\ 1 & 1-i & -i \\ i & i+1 & 1 \end{pmatrix}. \tag{13.30}$$

For matrices $\mathbf{A} \in \mathbb{C}^{M \times K}$, $\mathbf{B} \in \mathbb{C}^{K \times N}$, and $\mathbf{C} \in \mathbb{C}^{M \times N}$, the standard matrix product is given by

$$\mathbf{C} = \mathbf{A}\,\mathbf{B}$$

$$\{\mathbf{C}\}_{m,n} = \sum_{k} \mathbf{A}_{m,k}\,\mathbf{B}_{k,n}. \tag{13.31}$$

When multiplying these objects, it is often worth keeping in mind that the operation sums over indices that are next to each other (in this case, the index k). The transpose of the product of matrices reverses the order of the product with the transpose of the individual variables, as is indicated by

$$\mathbf{C}^{T} = (\mathbf{A}\,\mathbf{B})^{T}$$

$$= \mathbf{B}^{T}\,\mathbf{A}^{T}. \tag{13.32}$$

Similarly, the Hermitian conjugate is given by

$$\mathbf{C}^{H} = (\mathbf{A}\,\mathbf{B})^{H}$$

$$= \mathbf{B}^{H}\,\mathbf{A}^{H}. \tag{13.33}$$

13.2.1   Matrix Inverse

If a square matrix is invertible (and not all are), then the matrix inverse of $\mathbf{M}$ satisfies

$$\mathbf{M}^{-1}\,\mathbf{M} = \mathbf{M}\,\mathbf{M}^{-1} = \mathbf{I}, \tag{13.34}$$

where

$$\mathbf{I} = \mathrm{diag}\{1, 1, \cdots, 1\} \tag{13.35}$$

is the identity matrix given by a diagonal matrix of 1s. If the size is not clear from context, we use $\mathbf{I}_{m}$ for a matrix of size m. Upon first considering the concept of a matrix inverse, it can be a little confusing. Except for special cases, the inverse of the matrix is not the same as inverting the elements of the matrix, so, in general,

$$\begin{pmatrix} a & b \\ c & d \end{pmatrix}^{-1} \neq \begin{pmatrix} a^{-1} & b^{-1} \\ c^{-1} & d^{-1} \end{pmatrix}, \tag{13.36}$$

although the inverse of the identity matrix is its inverse. The inverse of the product of nonsingular square matrices (that is, invertible matrices) is given by

$$(\mathbf{A}\,\mathbf{B})^{-1} = \mathbf{B}^{-1}\,\mathbf{A}^{-1}. \tag{13.37}$$

We can explicitly express the inverse of a 2×2 matrix by

$$\begin{pmatrix} a & b \\ c & d \end{pmatrix}^{-1} = \frac{1}{ad - bc} \begin{pmatrix} d & -b \\ -c & a \end{pmatrix}. \tag{13.38}$$

The general inverse of a partitioned matrix is given by

$$\begin{pmatrix} \mathbf{A} & \mathbf{B} \\ \mathbf{C} & \mathbf{D} \end{pmatrix}^{-1} = \begin{pmatrix} (\mathbf{A} - \mathbf{B}\mathbf{D}^{-1}\mathbf{C})^{-1} & -\mathbf{A}^{-1}\mathbf{B}(\mathbf{D} - \mathbf{C}\mathbf{A}^{-1}\mathbf{B})^{-1} \\ -\mathbf{D}^{-1}\mathbf{C}(\mathbf{A} - \mathbf{B}\mathbf{D}^{-1}\mathbf{C})^{-1} & (\mathbf{D} - \mathbf{C}\mathbf{A}^{-1}\mathbf{B})^{-1} \end{pmatrix}. \tag{13.39}$$

13.2.2 Inversion of Matrix Sum

A general form of Woodbury's formula is given by

$$(\mathbf{M} + \mathbf{A}\mathbf{B})^{-1} = \mathbf{M}^{-1} - \mathbf{M}^{-1}\mathbf{A}(\mathbf{I} + \mathbf{B}\mathbf{M}^{-1}\mathbf{A})^{-1}\mathbf{B}\mathbf{M}^{-1}. \tag{13.40}$$

A special and useful form of Woodbury's formula is used to find the inverse of the identity matrix plus a rank-1 matrix:

$$(\mathbf{I} + \mathbf{v}\mathbf{w}^H)^{-1} = \mathbf{I} - \frac{\mathbf{v}\mathbf{w}^H}{1 + \mathbf{w}^H\mathbf{v}}. \tag{13.41}$$

The inverse of the identity matrix plus two rank-1 matrices is also useful. Here, the special case of a Hermitian matrix is considered. The matrix to be inverted is given by

$$\mathbf{I} + \mathbf{a}\mathbf{a}^H + \mathbf{b}\mathbf{b}^H, \tag{13.42}$$

where $\mathbf{a}$ and $\mathbf{b}$ are column vectors of the same size. The inverse is given by

$$(\mathbf{I} + \mathbf{a}\mathbf{a}^H + \mathbf{b}\mathbf{b}^H)^{-1} = \mathbf{I} - \left(\frac{\mathbf{a}\mathbf{a}^H}{1 + \mathbf{a}^H\mathbf{a}} + \frac{\mathbf{b}\mathbf{b}^H}{1 + \mathbf{b}^H\mathbf{b}} \right) \left(1 + \frac{\|\mathbf{a}^H\mathbf{b}\|^2}{\gamma} \right)$$
$$+ \frac{1}{\gamma} \left(\mathbf{a}^H\mathbf{b}\mathbf{a}\mathbf{b}^H + \mathbf{b}^H\mathbf{a}\mathbf{b}\mathbf{a}^H \right), \tag{13.43}$$

where, here,

$$\gamma = 1 + \mathbf{a}^H\mathbf{a} + \mathbf{b}^H\mathbf{b} + \mathbf{a}^H\mathbf{a}\mathbf{b}^H\mathbf{b} - \|\mathbf{a}^H\mathbf{b}\|^2. \tag{13.44}$$

This result can be found by employing Woodbury's formula with $\mathbf{M}$ in Equation (13.40):

$$\mathbf{M} = \mathbf{I} + \mathbf{b}\mathbf{b}^H$$

$$\mathbf{M}^{-1} = \mathbf{I} - \frac{\mathbf{b}\mathbf{b}^H}{1 + \mathbf{b}^H\mathbf{b}}. \tag{13.45}$$

Consequently, Woodbury's formula provides the form

$$(\mathbf{I} + \mathbf{a}\mathbf{a}^H + \mathbf{b}\mathbf{b}^H)^{-1} = (\mathbf{I} + \mathbf{b}\mathbf{b}^H)^{-1} \tag{13.46}$$
$$- (\mathbf{I} + \mathbf{b}\mathbf{b}^H)^{-1}\mathbf{a}(1 + \mathbf{a}^H[\mathbf{I} + \mathbf{b}\mathbf{b}^H]^{-1}\mathbf{a})^{-1}\mathbf{a}^H(\mathbf{I} + \mathbf{b}\mathbf{b}^H)^{-1},$$

which, after a bit of manipulation, is given by the form in Equation (13.43).

13.3   Special Matrices

13.3.1   Hermitian Matrix

There are a number of matrices that satisfy special symmetries. An important matrix type is the Hermitian matrix, which is a square matrix that satisfies

$$\mathbf{M}^H = \mathbf{M}. \tag{13.47}$$

One way in which we produce these Hermitian matrices is when we evaluate the product of a matrix, $\mathbf{Y}$, with its Hermitian conjugate,

$$\mathbf{M} = \mathbf{Y}\mathbf{Y}^H \tag{13.48}$$
$$\mathbf{M}^H = (\mathbf{Y}\mathbf{Y}^H)^H = \mathbf{Y}^{HH}\mathbf{Y}^H = \mathbf{Y}\mathbf{Y}^H.$$

A useful property of Hermitian matrices constructed from these quadratic forms is that they are positive-semidefinite. A positive-semidefinite matrix $\mathbf{M} \in \mathbb{C}^{m\times m}$ has the property that, for any nonzero vector $\mathbf{x} \in \mathbb{C}^{m\times 1}$, the following quadratic form is greater than or equal to zero:

$$\mathbf{x}^H\mathbf{M}\mathbf{x} = \mathbf{x}^H\mathbf{Y}\mathbf{Y}^H\mathbf{x}$$
$$\geq 0. \tag{13.49}$$

A related matrix is the positive-definite matrix that satisfies

$$\mathbf{x}^H\mathbf{M}\mathbf{x} > 0. \tag{13.50}$$

A positive-definite matrix is a nonsingular matrix, so it is invertible. Not all nonsingular matrices are positive-definite. The inverse, Hermitian, and transpose operations commute, so that

$$(\mathbf{M}^H)^{-1} = (\mathbf{M}^{-1})^H \text{ and } (\mathbf{M}^T)^{-1} = (\mathbf{M}^{-1})^T. \tag{13.51}$$

13.3.2   Unitary Matrix

An important type of matrix is the square matrix that preserves the length of a vector of dimension N when the matrix multiplies the vector. That is, $\mathbf{U}\mathbf{v}$ is the same length as $\mathbf{v}$. One can think of this as "rotating" the vector, but this could be in a complex high-dimensional space that is difficult to visualize.

The unitary matrix also has the interesting property that its inverse is the Hermitian conjugate. This can be particularly valuable because evaluating matrix inverses can be computationally expensive, while performing the Hermitian conjugation is easy. A unitary matrix is a square matrix that satisfies

$$\mathbf{U}^H = \mathbf{U}^{-1} \text{ so that } \mathbf{U}^H\mathbf{U} = \mathbf{U}\mathbf{U}^H = \mathbf{I}. \tag{13.52}$$

All the columns are orthonormal to all the other columns. All the rows are orthonormal to all the other rows. Thus, for any two columns in $\mathbf{U}$, which we denote by $\mathbf{u}_p$ and $\mathbf{u}_q$, the Hermitian inner product $\mathbf{u}_p^H\mathbf{u}_q = \delta_{p,q}$, where $\delta_{p,q}$ is the Kronecker delta.

The length l of a vector $\mathbf{v}$ in an L2-norm sense (which we discuss in Section 13.4.2) is given by

$$l = \sqrt{\sum_{m=1}^{N}\{\mathbf{v}\}_m^2} = \sqrt{\mathbf{v}^H\,\mathbf{v}} \tag{13.53}$$

$$\tilde{l} = \sqrt{\tilde{\mathbf{v}}^H\,\tilde{\mathbf{v}}}\,; \quad \tilde{\mathbf{v}} = \mathbf{U}\,\mathbf{v}$$

$$= \sqrt{\sum_{m=1}^{N}\{\mathbf{U}\,\mathbf{v}\}_m^2}$$

$$= \sqrt{\sum_{m=1}^{N}\{\mathbf{U}\,\mathbf{v}\}_m^*\,\{\mathbf{U}\,\mathbf{v}\}_m}\,;$$

$$= \sqrt{\sum_{m=1}^{N}\{(\mathbf{U}\,\mathbf{v})^H\}_m\,\{\mathbf{U}\,\mathbf{v}\}_m}$$

$$= \sqrt{\mathbf{v}^H\,\mathbf{U}^H\,\mathbf{U}\,\mathbf{v}}$$

$$= \sqrt{\mathbf{v}^H\,\mathbf{v}} = l \tag{13.54}$$

if $\mathbf{U}$ is unitary. The unitary matrix satisfies this length-preservation concept for all $\mathbf{v}$. A trivial example of a unitary matrix $\mathbf{U}$ is given by

$$\mathbf{U} = \frac{1}{\sqrt{2}}\begin{pmatrix} 1 & -1 \\ 1 & 1 \end{pmatrix}. \tag{13.55}$$

The Hermitian conjugate is then given by

$$\mathbf{U}^H = \frac{1}{\sqrt{2}}\begin{pmatrix} 1 & 1 \\ -1 & 1 \end{pmatrix} \tag{13.56}$$

so that

$$\mathbf{U}\,\mathbf{U}^H = \frac{1}{\sqrt{2}}\begin{pmatrix} 1 & -1 \\ 1 & 1 \end{pmatrix}\frac{1}{\sqrt{2}}\begin{pmatrix} 1 & 1 \\ -1 & 1 \end{pmatrix} = \begin{pmatrix} 1 & 0 \\ 0 & 1 \end{pmatrix}. \tag{13.57}$$

13.3.3 Element-Shifted Symmetries

Toeplitz matrices are of particular interest because not only are they produced in certain physical examples, they have fast inversion algorithms. While it takes order n^3 operations for a general square matrix of size n, a Toeplitz matrix can be inverted in order n^2 operations [28].

An $n \times n$ Toeplitz matrix is a matrix in which the values are equal along diagonals:

$$
\mathbf{M} =
\begin{pmatrix}
a_0 & a_{-1} & a_{-2} & \cdots & a_{-n+1} \\
a_1 & a_0 & a_{-1} & & a_{-n+2} \\
a_2 & a_1 & a_0 & & \\
\vdots & & & \ddots & a_{-1} \\
a_{n-1} & a_{n-2} & & a_1 & a_0
\end{pmatrix} .
\tag{13.58}
$$

The Toeplitz matrix is defined by $2n - 1$ values. An $n \times n$ circulant matrix is a special form of a Toeplitz matrix in which each row or column is a cyclic permutation of the previous row or column:

$$
\mathbf{M} =
\begin{pmatrix}
a_0 & a_{n-1} & a_{n-2} & \cdots & a_1 \\
a_1 & a_0 & a_{n-1} & & a_2 \\
a_2 & a_1 & a_0 & & \\
\vdots & & & \ddots & a_{n-1} \\
a_{n-1} & a_{n-2} & & a_1 & a_0
\end{pmatrix} .
\tag{13.59}
$$

The circulant matrix is defined by n values. An additional property of circulant matrices is that they can be inverted in the order of $n \log n$ operations.

13.4   Norms, Traces, and Determinants

When working with vectors and matrices, it is often useful to work with metrics that provide some characteristic metric of the object. Many of the relationships associated with norms, traces, and determinants become easier to demonstrate after the discussion of matrix decomposition found in Section 13.6.

13.4.1   Trace

The trace of a square matrix $\mathbf{M} \in \mathbb{C}^{m \times m}$ of size m is the sum of its diagonal elements. We indicate the trace by

$$
\text{tr}\{\mathbf{M}\} = \sum_m (\mathbf{M})_{m,m} .
\tag{13.60}
$$

The product of two matrices commutes under the trace operation:

$$
\text{tr}\{\mathbf{A}\,\mathbf{B}\} = \text{tr}\{\mathbf{B}\,\mathbf{A}\} .
\tag{13.61}
$$

We can motivate this relationship by reminding ourselves that the $\{\mathbf{A}\,\mathbf{B}\}_{m,n}$ is given by

$$
\{\mathbf{A}\,\mathbf{B}\}_{m,n} = \sum_k \{\mathbf{A}\}_{m,k}\,\{\mathbf{B}\}_{k,n} ,
\tag{13.62}
$$

so the trace is then given by

$$\text{tr}\{\mathbf{A}\,\mathbf{B}\} = \sum_m \{\mathbf{A}\,\mathbf{B}\}_{m,m}$$

$$= \sum_{m,k} \{\mathbf{A}\}_{m,k}\,\{\mathbf{B}\}_{k,m} = \sum_{m,k} \{\mathbf{B}\}_{k,m}\,\{\mathbf{A}\}_{m,k}$$

$$= \sum_k \{\mathbf{B}\,\mathbf{A}\}_{k,k} = \text{tr}\{\mathbf{B}\,\mathbf{A}\}. \tag{13.63}$$

We see from the relationship that the order of the matrices doesn't matter. This property can be extended to the product of three (or more) matrices such that

$$\text{tr}\{\mathbf{A}\,\mathbf{B}\,\mathbf{C}\} = \text{tr}\{\mathbf{C}\,\mathbf{A}\,\mathbf{B}\} = \text{tr}\{\mathbf{B}\,\mathbf{C}\,\mathbf{A}\}. \tag{13.64}$$

An interesting comment on the trace of a matrix is that if you rotate the axes so that all the elements of the matrix change, the trace remains the same. For example, if we perform a unitary transformation of some matrix $\mathbf{M}$ with unitary matrix $\mathbf{U}$, such that we replace $\mathbf{M}$ with $\mathbf{U}\,\mathbf{M}\,\mathbf{U}^H$, then the transformation does not affect the trace. To show this preservation, we use the commutation properties of the trace and the inversion properties of the unitary matrix, which gives

$$\text{tr}\{\mathbf{U}\,\mathbf{M}\,\mathbf{U}^H\} = \text{tr}\{\mathbf{U}^H\,\mathbf{U}\,\mathbf{M}\} = \text{tr}\{\mathbf{U}^{-1}\,\mathbf{U}\,\mathbf{M}\} = \text{tr}\{\mathbf{M}\}. \tag{13.65}$$

13.4.2 Norm

The norm is used to express some sort of length of a vector. The length we usually think of is given by the L2-norm. We indicate the absolute value of a scalar and the L2-norm of a vector by $\|\cdot\|$. On occasion, we may also use $Abs(\cdot)$ to indicate the absolute value of a scalar. For the sake of clarity, we reserve the notation $|.|$ exclusively for the determinant of a matrix. The absolute value of a scalar a is thus $\|a\|$, and the L2-norm of a vector $\mathbf{a}$ is denoted as follows:

$$\|\mathbf{a}\| = \sqrt{\sum_m \|(\mathbf{a})_m\|^2} = \sqrt{\mathbf{a}^H\,\mathbf{a}}. \tag{13.66}$$

To make contact with our comments on unitary matrices in Section 13.3.2, we note that the L2-norm of a vector and the L2-norm unitary transformation of the vector are the same, so for some unitary matrix $\mathbf{U}$,

$$\|\mathbf{U}\,\mathbf{a}\| = \|\mathbf{a}\|. \tag{13.67}$$

For the sake of generality, we can extend this idea to include other powers. We indicate the p-norm of a vector for values other than 2 by

$$\|\mathbf{a}\|_p = \left(\sum_m \|(\mathbf{a})_m\|^p\right)^{1/p}. \tag{13.68}$$

A couple of less than obvious norms are the L0-norm $\|\mathbf{a}\|_0$, which counts nonzero entries in the vector, and the infinity norm $\|\mathbf{a}\|_\infty$, which finds the entry with the maximum magnitude.

We can also extend this idea to include matrices by stacking up all the columns of a matrix into one long vector. If we take the L2-norm of that vector, we get the Frobenius norm of a matrix. We indicate the Frobenius norm of a matrix by

$$\|\mathbf{M}\|_F = \|\text{vec}\{\mathbf{M}\}\| = \sqrt{\sum_{m,n} \|(\mathbf{M})_{m,n}\|^2} = \sqrt{\text{tr}\{\mathbf{M}\,\mathbf{M}^H\}}\,. \tag{13.69}$$

Interestingly, the Frobenius norm of a matrix can also be evaluated by using the square root of the trace of the product of the matrix with the Hermitian conjugate of the matrix. We can think of the square Frobenius norm of a matrix as the total power or energy of the matrix.

13.4.3   Determinants

The determinant has a nice geometric interpretation. It is the volume of the parallelepiped spanned by column vectors of the matrix. We can evaluate iteratively the determinant of the matrix $\mathbf{A}$ by

$$|\mathbf{A}| = \sum_n (\mathbf{A})_{m,n}\,(-1)^{m+n}\,|\mathbf{M}_{m,n}|\,, \tag{13.70}$$

where submatrix $\mathbf{M}_{m,n}$ is here defined to be the minor of $\mathbf{A}$, which is constructed by removing the mth row and nth column of $\mathbf{A}$ (not to be confused with the mth, nth element of $\mathbf{A}$ or $\mathbf{M}$). The determinant of a 2×2 matrix is given by

$$\left|\begin{pmatrix} a & b \\ c & d \end{pmatrix}\right| = ad - bc\,. \tag{13.71}$$

The determinant has a number of useful relationships. The determinant of the product of square matrices is equal to the product of the matrix determinants,

$$|\mathbf{A}\,\mathbf{B}| = |\mathbf{A}|\,|\mathbf{B}| = |\mathbf{B}\,\mathbf{A}|\,. \tag{13.72}$$

For some scalar $c \in \mathbb{C}$ and matrix $\mathbf{M} \in \mathbb{C}^{m \times m}$, the determinant of the product is the product of the scalar to the mth power times the determinant of the matrix,

$$|c\,\mathbf{M}| = c^m\,|\mathbf{M}|\,. \tag{13.73}$$

The determinant of the identity matrix is given by

$$|\mathbf{I}| = 1\,, \tag{13.74}$$

and the determinant of a unitary matrix $\mathbf{U}$ is magnitude 1:

$$\|\,|\mathbf{U}|\,\| = 1\,. \tag{13.75}$$

Consequently, the determinant of the unitary transformation with unitary matrix $\mathbf{U}$, defined in Equation (13.52), of a matrix $\mathbf{A}$ is the determinant of $\mathbf{A}$:

$$|\mathbf{U}\,\mathbf{A}\,\mathbf{U}^H| = |\mathbf{U}\,\mathbf{A}|\,|\mathbf{U}^H|$$
$$= |\mathbf{U}^H\,\mathbf{U}\,\mathbf{A}|$$
$$= |\mathbf{A}|\,. \tag{13.76}$$

The product of matrices plus the identity matrix commute under the determinant so that

$$|\mathbf{I} + \mathbf{A}\,\mathbf{B}| = |\mathbf{I} + \mathbf{B}\,\mathbf{A}|, \tag{13.77}$$

where $\mathbf{A} \in \mathbb{C}^{m \times n}$ and $\mathbf{B} \in \mathbb{C}^{n \times m}$ are not necessarily square (although $\mathbf{AB}$ and $\mathbf{BA}$ are). The inverse of a matrix determinant is equal to the determinant of a matrix inverse:

$$|\mathbf{M}|^{-1} = |\mathbf{M}^{-1}|. \tag{13.78}$$

13.5 Special Vector and Matrix Products

In addition to the standard matrix and vector products that we defined previously, a few more are useful. Maybe amusingly, the Hadamard product is probably what you would naively expect a matrix or vector product to be. For matrices $\mathbf{A} \in \mathbb{C}^{M \times N}$, $\mathbf{B} \in \mathbb{C}^{M \times N}$, and $\mathbf{C} \in \mathbb{C}^{M \times N}$, the Hadamard or element-by-element product is denoted by $\cdot \odot \cdot$ such that

$$\mathbf{C} = \mathbf{A} \odot \mathbf{B}$$
$$\{\mathbf{C}\}_{m,n} = \{\mathbf{A} \odot \mathbf{B}\}_{m,n}$$
$$= \mathbf{A}_{m,n}\,\mathbf{B}_{m,n}. \tag{13.79}$$

For matrices $\mathbf{A} \in \mathbb{C}^{M \times N}$, $\mathbf{B} \in \mathbb{C}^{J \times K}$, and $\mathbf{C} \in \mathbb{C}^{MJ \times NK}$, the standard definition of the Kronecker product [31] is denoted by $\cdot \otimes \cdot$ and is given by

$$\mathbf{C} = \mathbf{A} \otimes \mathbf{B}$$
$$= \begin{pmatrix} \{\mathbf{A}\}_{1,1}\,\mathbf{B} & \{\mathbf{A}\}_{1,2}\,\mathbf{B} & \{\mathbf{A}\}_{1,3}\,\mathbf{B} & \cdots \\ \{\mathbf{A}\}_{2,1}\,\mathbf{B} & \{\mathbf{A}\}_{2,2}\,\mathbf{B} & \{\mathbf{A}\}_{2,3}\,\mathbf{B} \\ \{\mathbf{A}\}_{3,1}\,\mathbf{B} & \{\mathbf{A}\}_{3,2}\,\mathbf{B} \\ \vdots \end{pmatrix}. \tag{13.80}$$

A few useful relationships are given here:

$$(\mathbf{A} \otimes \mathbf{B})^T = \mathbf{A}^T \otimes \mathbf{B}^T \tag{13.81}$$
$$(\mathbf{A} \otimes \mathbf{B})^* = \mathbf{A}^* \otimes \mathbf{B}^* \tag{13.82}$$
$$(\mathbf{A} \otimes \mathbf{B})^H = \mathbf{A}^H \otimes \mathbf{B}^H \tag{13.83}$$
$$(\mathbf{A} \otimes \mathbf{B})^{-1} = \mathbf{A}^{-1} \otimes \mathbf{B}^{-1}, \tag{13.84}$$

where it is assumed that $\mathbf{A}$ and $\mathbf{B}$ are not singular for the last relationship. The Kronecker product obeys distributive and associative properties,

$$(\mathbf{A} + \mathbf{B}) \otimes \mathbf{C} = \mathbf{A} \otimes \mathbf{C} + \mathbf{B} \otimes \mathbf{C} \tag{13.85}$$
$$(\mathbf{A} \otimes \mathbf{B}) \otimes \mathbf{C} = \mathbf{A} \otimes (\mathbf{B} \otimes \mathbf{C}). \tag{13.86}$$

The product of Kronecker products is given by

$$(\mathbf{A} \otimes \mathbf{B})(\mathbf{C} \otimes \mathbf{D}) = (\mathbf{AC}) \otimes (\mathbf{BD}). \tag{13.87}$$

For square matrices $\mathbf{A} \in \mathbb{C}^{M \times M}$ and $\mathbf{B} \in \mathbb{C}^{N \times N}$,

$$\text{tr}\{\mathbf{A} \otimes \mathbf{B}\} = \text{tr}\{\mathbf{A}\}\, \text{tr}\{\mathbf{B}\} \tag{13.88}$$

$$|\mathbf{A} \otimes \mathbf{B}| = |\mathbf{A}|^N\, |\mathbf{B}|^M, \tag{13.89}$$

where trace and determinants are defined in Section 13.4. Note that the exponents M and N are for the size of the opposing matrix. The vector operation and Kronecker product are related by

$$\text{vec}(\mathbf{a}\, \mathbf{b}^T) = \mathbf{b} \otimes \mathbf{a} \tag{13.90}$$

$$\text{vec}(\mathbf{A}\, \mathbf{B}\, \mathbf{C}) = (\mathbf{C}^T \otimes \mathbf{A})\, \text{vec}(\mathbf{B}). \tag{13.91}$$

If the dimensions of $\mathbf{A}$ and $\mathbf{B}$ are the same and the dimensions of $\mathbf{C}$ and $\mathbf{D}$ are the same, then Hadamard and Kronecker products are related by

$$(\mathbf{A} \odot \mathbf{B}) \otimes (\mathbf{C} \odot \mathbf{D}) = (\mathbf{A} \otimes \mathbf{C}) \odot (\mathbf{B} \otimes \mathbf{D}). \tag{13.92}$$

13.6 Matrix Decompositions

13.6.1 Eigen Analysis

One of the essential tools for signal processing is the eigenvalue decomposition. A complex square matrix $\mathbf{M} \in \mathbb{C}^{M \times M}$ has M eigenvalues and eigenvectors. The mth eigenvalue λ_m – based on some metric for ordering such as magnitude – and the corresponding mth eigenvector $\mathbf{v}_m$ of the matrix $\mathbf{M}$ are given by the solution of

$$\mathbf{M}\, \mathbf{v}_m = \lambda_m \mathbf{v}_m. \tag{13.93}$$

Sometimes, for clarity, we indicate the mth eigenvalue of a matrix $\mathbf{M}$ by $\lambda_m\{\mathbf{M}\}$. A matrix is classified as positive-definite if all the eigenvalues are real and positive ($\lambda_m\{\mathbf{M}\} > 0 \,\forall\, m \in \{1, \cdots, M\}$) and is classified as positive-semidefinite if all the eigenvalues are positive or zero ($\lambda_m\{\mathbf{M}\} \geq 0 \,\forall\, m \in \{1, \cdots, M\}$). In cases in which there are duplicated or degenerate eigenvalues, the eigenvectors are only determined to within a subspace of dimension of the number of degenerate eigenvalues. Any vector from an orthonormal basis in that subspace would satisfy Equation (13.93). The extreme example is the identity matrix in which all the eigenvalues are the same. In this case, the subspace is the entire space; thus, any vector satisfies Equation (13.93). As an example, consider the matrix $\mathbf{M}$ that is constructed by the quadratic form using $\mathbf{X}$, given by

$$\mathbf{M} = \mathbf{X}\mathbf{X}^H$$

$$= \begin{pmatrix} 3 & 4 & -3 \\ 1 & -2 & -3 \\ 0 & -2 & 4 \end{pmatrix} \begin{pmatrix} 3 & 4 & -3 \\ 1 & -2 & -3 \\ 0 & -2 & 4 \end{pmatrix}^H$$

$$= \begin{pmatrix} 34 & 4 & -20 \\ 4 & 14 & -8 \\ -20 & -8 & 20 \end{pmatrix} \tag{13.94}$$

$$\lambda_1 \approx 49.9, \quad \lambda_2 \approx 14.3, \quad \lambda_3 \approx 3.8, \tag{13.95}$$

which we plot in Figure 13.2.

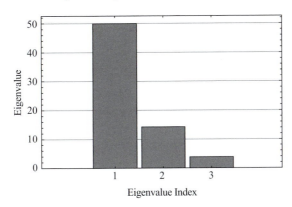

Figure 13.2 Example of sorted eigenvalues for **M** given in Equation (13.94).

We indicate the sum of the diagonal elements of a matrix by the trace. The trace is also equal to the sum of the eigenvalues of the matrix:

$$\text{tr}\{\mathbf{M}\} = \sum_m (\mathbf{M})_{m,m} = \sum_m \lambda_m . \tag{13.96}$$

The determinant of a matrix is equal to the product of its eigenvalues:

$$|\mathbf{M}| = \prod_m \lambda_m . \tag{13.97}$$

While, in general, for some square matrices **A** and **B**, the mth eigenvalue of the matrix sum does not equal the sum of the mth eigenvalues of the individual matrices:

$$\lambda_m\{\mathbf{A} + \mathbf{B}\} \neq \lambda_m\{\mathbf{A}\} + \lambda_m\{\mathbf{B}\} , \tag{13.98}$$

for the special case of $\mathbf{I} + \mathbf{A}$, the eigenvalues add as follows:

$$\lambda_m\{\mathbf{I} + \mathbf{A}\} = 1 + \lambda_m\{\mathbf{A}\} . \tag{13.99}$$

As an example, consider the matrix **M** that is constructed by the quadratic form using **v** plus the identity matrix. The form $\mathbf{v}\mathbf{v}^H$ produces a rank-1 matrix that only has a single nonzero eigenvalue (discussed in Section 13.6.3). The eigenvalues for this matrix are given by

$$\mathbf{M} = \mathbf{I} + \mathbf{v}\mathbf{v}^H$$

$$= \mathbf{I} + \begin{pmatrix} 2 \\ -3 \\ 1 \end{pmatrix} \begin{pmatrix} 2 \\ -3 \\ 1 \end{pmatrix}^H$$

$$= \begin{pmatrix} 5 & -6 & 2 \\ -6 & 10 & -3 \\ 2 & -3 & 2 \end{pmatrix} \tag{13.100}$$

$$\lambda_1 = 15, \quad \lambda_2 = 1, \quad \lambda_3 = 1, \tag{13.101}$$

which we plot in Figure 13.3.

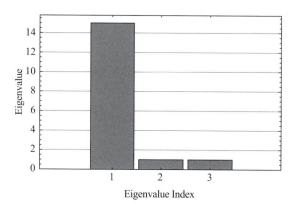

Figure 13.3 Example of sorted eigenvalues for **M** given in Equation (13.100).

13.6.2 Eigenvalues of 2×2 Hermitian Matrix

Given the 2×2 Hermitian matrix $\mathbf{M} \in \mathbb{C}^{2\times2}$ (which is a matrix that satisfies $\mathbf{M} = \mathbf{M}^H$), where

$$\mathbf{M} = \begin{pmatrix} a & c^* \\ c & b \end{pmatrix}, \tag{13.102}$$

the eigenvalues of $\mathbf{M}$ can be found by exploiting the knowledge of the eigenvalue relationships between the trace and the determinant. Because the matrix is Hermitian, the diagonal values (a and b) are real. The trace of $\mathbf{M}$ is given by

$$\text{tr}\{\mathbf{M}\} = \lambda_1 + \lambda_2$$
$$= a + b. \tag{13.103}$$

The determinant of the Hermitian matrix $\mathbf{M}$ is given by

$$|\mathbf{M}| = \lambda_1 \lambda_2$$
$$= ab - \|c\|^2. \tag{13.104}$$

By combining these two results, the eigenvalues can be explicitly found. The eigenvalues are given by

$$\lambda_1 + \lambda_2 = a + b$$
$$\lambda_1^2 + \lambda_1\lambda_2 = (a + b)\lambda_1$$
$$\lambda_1^2 + ab - \|c\|^2 = (a + b)\lambda_1 ; \quad \lambda_1 \lambda_2 = ab - \|c\|^2$$
$$0 = \lambda_1^2 - (a + b)\lambda_1 + ab - \|c\|^2. \tag{13.105}$$

Because the calculation is symmetric under swapping λ_1 and λ_2, we just solve for the two possible solutions of λ given by

$$\lambda = \frac{a+b \pm \sqrt{(a+b)^2 - 4(ab - \|c\|^2)}}{2}$$

$$= \frac{a+b \pm \sqrt{(a-b)^2 + 4\|c\|^2}}{2}. \tag{13.106}$$

As will be discussed in Section 13.6.4, Hermitian matrices constructed from quadratic forms are positive-semidefinite.

13.6.3   Eigenvalues of Low-Rank Matrices

We define a low-rank matrix to be a matrix for which some (and usually most) of the eigenvalues are zero. In a variety of applications, such as spatial covariance matrices, low-rank matrices are constructed with the outer product of vectors.

Rank-1 Matrix

For example, we define a rank-1 square matrix $\mathbf{M}$ to be one that we construct by using complex n-vectors $\mathbf{v} \in \mathbb{C}^{n \times 1}$ and $\mathbf{w} \in \mathbb{C}^{n \times 1}$,

$$\mathbf{M} = \mathbf{v}\mathbf{w}^H. \tag{13.107}$$

This matrix has an eigenvector proportional to $\mathbf{v}$ and an eigenvalue of $\mathbf{w}^H \mathbf{v}$. The eigenvalue can be determined directly by noting that the trace of the matrix is equal to the sum of the eigenvalues, which (for a rank-1 matrix) are all zero except for one. For comparison, if we consider the singular value decomposition (SVD) that we discuss in Section 13.6.4, this matrix has a single nonzero singular value given by $\|\mathbf{w}\| \, \|\mathbf{v}\|$.

Rank-2 Matrix

A Hermitian rank-2 matrix $\mathbf{M}$ can be constructed by using the two n-vectors $\mathbf{x} \in \mathbb{C}^{n \times 1}$ and $\mathbf{y} \in \mathbb{C}^{n \times 1}$ such that

$$\mathbf{M} = \mathbf{x}\mathbf{x}^H + \mathbf{y}\mathbf{y}^H. \tag{13.108}$$

The eigenvalues can be found by using the hypothesis that the eigenvector is proportional to $\mathbf{x} + a\mathbf{y}$, where a is some undetermined constant. The nonzero eigenvalues of $\mathbf{M}$ are given by λ_+ and λ_-:

$$\lambda_{\pm}\{\mathbf{M}\} = \frac{\|\mathbf{x}\|^2 + \|\mathbf{y}\|^2 \pm \sqrt{\left(\|\mathbf{x}\|^2 - \|\mathbf{y}\|^2\right)^2 + 4\|\mathbf{x}^H\mathbf{y}\|^2}}{2}. \tag{13.109}$$

13.6.4   Singular Value Decomposition

Another important concept is the SVD. The SVD of a matrix decomposes a matrix $\mathbf{A}$ into three matrices: an orthonormal matrix, a diagonal matrix containing the singular values, and another orthonormal matrix,

$$A = U S V^H. \tag{13.110}$$

The columns of the orthonormal matrices are orthogonal and are unit norm such that the vector $\mathbf{u}_m$ constructed from the mth column satisfies $\mathbf{u}_m^H \mathbf{u}_n = \delta_{m,n}$, where $\delta_{m,n}$ is the Kronecker delta. If the matrix is square, then the orthonormal matrices become unitary matrices so that the decomposition is given by a unitary matrix, a diagonal matrix containing the singular values, and another unitary matrix, where $\mathbf{U}$ and $\mathbf{V}$ are then unitary matrices and the diagonal matrix

$$\mathbf{S} = \begin{pmatrix} s_1 & 0 & 0 & \cdots \\ 0 & s_2 & 0 & \\ 0 & 0 & s_3 & \\ \vdots & & & \ddots \end{pmatrix} \tag{13.111}$$

contains the singular values $s_1, s_2, \cdots$. In the decomposition, there is sufficient freedom to impose the requirement that the singular values are real and positive. Note that the singular matrix $\mathbf{S}$ need not be square. In fact, the dimensions of $\mathbf{S}$ are the same as the dimensions of $\mathbf{A}$ since both the right and left singular matrices $\mathbf{U}$ and $\mathbf{V}$ are square. The mth column in either $\mathbf{U}$ or $\mathbf{V}$ is said to be the mth left-hand or right-hand singular vectors associated with the mth singular value, s_m.

The rank of some matrix $\mathbf{M}$ is the size of the vector space spanned by the matrix's columns. We can express this as the number of nonzero singular values:

$$\text{rank}\{\mathbf{M}\} = \#\{m : s_m \neq 0\}, \tag{13.112}$$

where we use $\#\{\cdot\}$ to indicate the number of entries that satisfy the condition. If a square matrix is full rank, meaning that the $\text{rank}\{\mathbf{M}\} = M$ for some matrix $\mathbf{M} \in \mathbb{C}^{M \times M}$ – by considering the SVD of a matrix – we can express the matrix's inverse with

$$\begin{aligned} \mathbf{M}^{-1} &= \left(\mathbf{U} \mathbf{D} \mathbf{V}^H\right)^{-1} \\ &= \mathbf{V} \mathbf{D}^{-1} \mathbf{U}^H, \end{aligned} \tag{13.113}$$

taking advantage of the fact that a Hermitian conjugate of a unitary matrix is its inverse, that is, $\mathbf{U}^H = \mathbf{U}^{-1}$.

The eigenvalues of the quadratic Hermitian form $\mathbf{A}\mathbf{A}^H$ are equal to the square of the singular values of $\mathbf{A}$:

$$\mathbf{A}\mathbf{A}^H = \mathbf{U} \mathbf{S} \mathbf{V}^H \mathbf{V} \mathbf{S}^H \mathbf{U}^H = \mathbf{U} \mathbf{S} \mathbf{S}^H \mathbf{U}^H, \tag{13.114}$$

where $\mathbf{S}\mathbf{S}^H = \text{diag}\{\|s_1\|^2, \|s_2\|^2, \cdots\}$. The columns of $\mathbf{U}$ are the eigenvectors of $\mathbf{A}\mathbf{A}^H$. The eigenvalues of a Hermitian form $\mathbf{A}\mathbf{A}^H$ are greater than or equal to zero, and thus the form $\mathbf{A}\mathbf{A}^H$ is said to be positive-semidefinite,

$$\lambda_m\{\mathbf{A}\mathbf{A}^H\} = (\mathbf{S}\mathbf{S}^H)_{m,m} \geq 0. \tag{13.115}$$

Notationally, a matrix with all-positive eigenvalues is said to be positive-definite, as defined in Equation (13.50). We indicate this by

$$\mathbf{M} > \mathbf{0} \rightarrow \lambda_m\{\mathbf{M}\} > 0 \quad \forall m. \tag{13.116}$$

We indicate a positive-semidefinite matrix, as defined in Equation (13.49), by

$$\mathbf{M} \geq 0 \rightarrow \lambda_m\{\mathbf{M}\} \geq 0 \quad \forall m. \tag{13.117}$$

13.6.5   QR Decomposition

Another common matrix decomposition is the QR factorization. In this decomposition, we factor a matrix $\mathbf{M}$ into a unitary matrix $\mathbf{Q}$ and an upper right-hand triangular matrix $\mathbf{R}$, where an upper right-hand triangular matrix has the form

$$\mathbf{R} = \begin{pmatrix} r_{1,1} & r_{1,2} & r_{1,3} & \cdots & r_{1,n} \\ 0 & r_{2,2} & r_{2,3} & \cdots & r_{2,n} \\ 0 & 0 & r_{3,3} & \cdots & r_{3,n} \\ \vdots & & & \ddots & \\ 0 & 0 & 0 & \cdots & r_{n,n} \end{pmatrix}. \tag{13.118}$$

For our purposes, we have an unfortunate naming convention problem because we often use $\mathbf{R}$ to indicate a covariance matrix rather than an upper triangle matrix. We hope that the correct meaning of $\mathbf{R}$ is clear from the context of its usage. If the matrix $\mathbf{M}$ has symmetric dimensions $n \times n$, then the decomposition is given by

$$\mathbf{M} = \mathbf{Q}\mathbf{R}. \tag{13.119}$$

For a rectangular matrix $\mathbf{M} \in \mathbb{C}^{m \times n}$ with $m > n$, the QR decomposition can be constructed so that

$$\mathbf{M} = \mathbf{Q} \begin{pmatrix} \mathbf{R} \\ \mathbf{0} \end{pmatrix}, \tag{13.120}$$

where the upper triangular matrix has dimensions $\mathbf{R} \in \mathbb{C}^{n \times n}$ and the zero matrix $\mathbf{0}$ has dimensions $(m - n) \times n$. As an example, consider the carefully selected matrix $\mathbf{M}$, given by

$$\mathbf{M} = \begin{pmatrix} 1 & 4 & 6 \\ 1 & -2 & 1 \\ 1 & 4 & 2 \\ 1 & -2 & -3 \end{pmatrix}. \tag{13.121}$$

The QR decomposition is typically evaluated numerically. A common approach to evaluate the QR is by implementing Householder transformations, although the details are beyond the scope of this text. Importantly, there are approaches to efficiently make these techniques to hardware implementations, which makes this decomposition useful. For the example matrix, the decomposition is given by

$$\mathbf{M} = \mathbf{Q}\mathbf{R} \tag{13.122}$$

$$\mathbf{Q} = \frac{1}{2} \begin{pmatrix} 1 & 1 & 1 & 1 \\ 1 & -1 & 1 & -1 \\ 1 & 1 & -1 & -1 \\ 1 & -1 & -1 & 1 \end{pmatrix} \tag{13.123}$$

$$\mathbf{R} = \begin{pmatrix} 2 & 2 & 3 \\ 0 & 6 & 5 \\ 0 & 0 & 4 \\ 0 & 0 & 0 \end{pmatrix}. \tag{13.124}$$

As we required, we observe $\mathbf{Q}\mathbf{Q}^H = \mathbf{I}$.

As an example application [32], for a square, invertible matrix $\mathbf{M}$, the inverse of $\mathbf{M}$ is given by

$$\mathbf{M}^{-1} = (\mathbf{Q}\mathbf{R})^{-1}$$
$$= \mathbf{R}^{-1}\mathbf{Q}^{-1}. \tag{13.125}$$

However, for vectors $\mathbf{w}$ and $\mathbf{v} \in \mathbb{C}^{N \times 1}$ and matrix $\mathbf{M} \in \mathbb{C}^{N \times N}$, we often have problems of the form

$$\mathbf{w} = \mathbf{M}^{-1}\mathbf{v}$$
$$\mathbf{M}\mathbf{w} = \mathbf{v} \tag{13.126}$$
$$\mathbf{Q}\mathbf{R}\mathbf{w} = \mathbf{v}$$
$$\mathbf{R}\mathbf{w} = \mathbf{Q}^{-1}\mathbf{v} = \mathbf{Q}^H\mathbf{v}, \tag{13.127}$$

where we have used the inversion characteristic of unitary matrices: $\mathbf{Q}^{-1} = \mathbf{Q}^H$. The form for $\mathbf{w}$ can be found by using the back-substitution algorithm for upper triangular matrices. The mth element of $\mathbf{w}$, denoted $\{\mathbf{w}\}_m$, is given by

$$\{\mathbf{w}\}_m = \frac{\{\mathbf{v}\}_m \sum_{k=m+1}^{N} \mathbf{Q}_{k,m}^* \{\mathbf{w}\}_k}{\{\mathbf{Q}\}_{m,m}^*}, \tag{13.128}$$

where we start with $m = N$ and work toward smaller values of m.

13.6.6 Matrix Subspaces

If we consider a vector space, a subspace is a portion of that space. We define this subspace by using a linear basis contained within the larger vector space, that is, the subspace is given by a set of vectors with arbitrary scaling. For example, we define a two-dimensional subspace as all points given by $b_1\mathbf{a}_1 + b_2\mathbf{a}_2$ for an arbitrary pair of vectors $\mathbf{a}_1$ and $\mathbf{a}_2$ for all values of b_1 and b_2, where we have assumed that $\mathbf{a}_1$ and $\mathbf{a}_2$ are not pointed along the same direction. For this example, we say that the subspace is spanned by $\mathbf{a}_1$ and $\mathbf{a}_2$. We also associate the subspace with the projection operator constructed from the vectors that span the space. It is also often useful to identify a subspace with projection operator that is orthogonal to any part of the vector space not contained within the subspace. Depending upon the application, vector spaces can be constructed by using either column vectors or row vectors.

We can decompose a matrix $\mathbf{M} \in \mathbb{C}^{m \times n}$ into components that occupy orthogonal subspaces that can be denoted by the matrices $\mathbf{M}_A \in \mathbb{C}^{m \times n}$ and $\mathbf{M}_{A^\perp} \in \mathbb{C}^{m \times n}$ such that

$$\mathbf{M} = \mathbf{M}_A + \mathbf{M}_{A^\perp}. \tag{13.129}$$

The matrix $\mathbf{M_A}$ can be constructed by projecting $\mathbf{M}$ onto the subspace spanned by the columns of the matrix $\mathbf{A} \in \mathbb{C}^{m \times m'}$, whose number of columns is less than or equal to the number of rows, that is, $m' \leq m$. It is assumed here that $\mathbf{A}^H \mathbf{A}$ is invertible and that we are operating on the column space of the matrix $\mathbf{M}$, although there is an equivalent row-space formulation. We can construct a projection matrix or projection operator $\mathbf{P_A} \in \mathbb{C}^{m \times m}$, which is given by

$$\mathbf{P_A} = \mathbf{A} (\mathbf{A}^H \mathbf{A})^{-1} \mathbf{A}^H . \tag{13.130}$$

For some matrix of an appropriate dimension $\mathbf{B}$, this projection matrix operates on the column space of $\mathbf{B}$ by multiplying the operator by the matrix $\mathbf{P_A} \mathbf{B}$. In higher dimensions, these ideas are difficult to visualize.

As an aside, it is worth noting that projection matrices are idempotent, that is, $\mathbf{P_A} \mathbf{P_A} = \mathbf{P_A}$. The matrix $\mathbf{M_A}$, which is the projection of $\mathbf{M}$ onto the subspace spanned by the columns of $\mathbf{A}$, is given by

$$\mathbf{M_A} = \mathbf{P_A} \mathbf{M} . \tag{13.131}$$

The rank of $\mathbf{M_A}$ is bounded by the number of columns in $\mathbf{A}$:

$$\text{rank}\{\mathbf{M_A}\} \leq m' . \tag{13.132}$$

As examples, consider the following scenarios. For vectors $\mathbf{a}$ and $\mathbf{b}$ in some higher-dimensional space, we can construct a rank-1 matrix with

$$\mathbf{A} = \mathbf{a} \mathbf{b}^H . \tag{13.133}$$

This matrix only has a single nonzero singular value. We can build a rank-2 matrix with

$$\mathbf{A} = \mathbf{a}_1 \mathbf{b}_1^H + \mathbf{a}_2 \mathbf{b}_2^H . \tag{13.134}$$

If we assume $\mathbf{a}_1 \neq \mathbf{a}_2$ and $\mathbf{b}_1 \neq \mathbf{b}_2$, then the rank of $\mathbf{A}$ is 2. However, if $\mathbf{a}_1 = \mathbf{a}_2 = \mathbf{a}$, then

$$\begin{aligned} \mathbf{A} &= \mathbf{a} \mathbf{b}_1^H + \mathbf{a} \mathbf{b}_2^H \\ &= \mathbf{a} (\mathbf{b}_1^H + \mathbf{b}_2^H) = \mathbf{a} \mathbf{c}^H , \end{aligned} \tag{13.135}$$

where $\mathbf{c} = \mathbf{b}_1 + \mathbf{b}_2$ so that this form produces a rank-1 matrix. Consequently, we can increase the rank until we reach the limit of the number of independent rows or columns in $\mathbf{A}$. Because we asserted that $\mathbf{A}$ has more rows than columns, we are limited by the number of columns.

The orthogonal projection matrix $\mathbf{P_A^\perp}$ is given by

$$\begin{aligned} \mathbf{P_A^\perp} &= \mathbf{I} - \mathbf{P_A} \\ &= \mathbf{I} - \mathbf{A} (\mathbf{A}^H \mathbf{A})^{-1} \mathbf{A}^H . \end{aligned} \tag{13.136}$$

We define the matrix $\mathbf{M_{A^\perp}}$ to be the matrix projected onto the basis orthogonal to $\mathbf{A}$, $\mathbf{M_{A^\perp}} = \mathbf{P_A^\perp} \mathbf{M}$. Consequently, the matrix $\mathbf{M}$ can be decomposed into the matrices

$$M = IM$$
$$= (P_A + P_A^\perp) M$$
$$= M_A + M_{A^\perp} . \tag{13.137}$$

As an example that we can visualize, we consider a three-dimensional real space. We set the matrix $A \in \mathbb{R}^{3 \times 2}$ such that A is given by

$$A = (\ a_1 \quad a_2 \ ) , \tag{13.138}$$

where a_1 and a_2 are in $\mathbb{R}^{3 \times 1}$. If we consider some arbitrary vector in the three-space v, some portion is in the subspace spanned by A and some portion is in the subspace orthogonal to A. In Figure 13.4, we depict the three-space with a projection of v into subspaces associated with A and orthogonal to A.

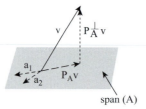

span (A)

Figure 13.4 Example of a vector v projected onto the subspaces defined by A and that which is orthogonal to A.

As we mentioned, there is also a row-space formulation. A projection matrix can project onto the row space of some matrix B,

$$B = \begin{pmatrix} \underline{b}_1 \\ \vdots \\ \underline{b}_n \end{pmatrix} , \tag{13.139}$$

where B is constructed from row vectors $\underline{b}_1, \cdots , \underline{b}_n$. To do this, we rearrange the projection matrix definition and operate from the right. For some matrix M projecting onto the row space of B, M_B is given by

$$M_B = M P_B \tag{13.140}$$
$$P_B = B^H (B B^H)^{-1} B . \tag{13.141}$$

13.7   Real Derivatives of Multivariate Expressions

The real derivatives of multivariate expressions follow directly from standard scalar derivatives [30]. For the real variable α, the derivatives of complex N-vector z and complex matrix M are given by

$$
\frac{\partial}{\partial \alpha} \mathbf{z} = \begin{pmatrix} \frac{\partial}{\partial \alpha} \{\mathbf{z}\}_1 \\ \frac{\partial}{\partial \alpha} \{\mathbf{z}\}_2 \\ \vdots \\ \frac{\partial}{\partial \alpha} \{\mathbf{z}\}_N \end{pmatrix} \tag{13.142}
$$

and

$$
\left\{ \frac{\partial}{\partial \alpha} \mathbf{M} \right\}_{m,n} = \frac{\partial}{\partial \alpha} \{\mathbf{M}\}_{m,n} , \tag{13.143}
$$

respectively.

A few useful expressions follow [30]. Under the assumption that the complex vectors $\mathbf{z}$ and $\mathbf{A}$ are functions of α, the derivative of the quadratic form $\mathbf{z}^H \mathbf{A} \mathbf{z}$ with respect to the real parameter α is given by

$$
\frac{\partial}{\partial \alpha} \mathbf{z}^H \mathbf{A} \mathbf{z} = \left(\frac{\partial}{\partial \alpha} \mathbf{z}^H \right) \mathbf{A} \mathbf{z} + \mathbf{z}^H \left(\frac{\partial}{\partial \alpha} \mathbf{A} \right) \mathbf{z} + \mathbf{z}^H \mathbf{A} \frac{\partial}{\partial \alpha} \mathbf{z} . \tag{13.144}
$$

The derivative for the complex invertible matrix $\mathbf{M}$ with respect to real parameter α can be found by considering the derivative of $\partial/\partial\alpha(\mathbf{M}\mathbf{M}^{-1}) = \partial/\partial\alpha\mathbf{I} = 0$ and is given by

$$
\frac{\partial}{\partial \alpha} \mathbf{M}^{-1} = -\mathbf{M}^{-1} \left(\frac{\partial}{\partial \alpha} \mathbf{M} \right) \mathbf{M}^{-1} . \tag{13.145}
$$

The derivatives of the determinant and the log determinant of a nonsingular matrix $\mathbf{M}$, with respect to real parameter α, are given by

$$
\frac{\partial}{\partial \alpha} |\mathbf{M}| = |\mathbf{M}| \, \mathrm{tr} \left\{ \mathbf{M}^{-1} \frac{\partial}{\partial \alpha} \mathbf{M} \right\} \tag{13.146}
$$

and

$$
\frac{\partial}{\partial \alpha} \log_e |\mathbf{M}| = \mathrm{tr} \left\{ \mathbf{M}^{-1} \frac{\partial}{\partial \alpha} \mathbf{M} \right\} , \tag{13.147}
$$

respectively. The derivative of the trace of a matrix is equal to the trace of the derivative of the matrix:

$$
\frac{\partial}{\partial \alpha} \mathrm{tr}\{\mathbf{M}\} = \mathrm{tr} \left\{ \frac{\partial}{\partial \alpha} \mathbf{M} \right\} . \tag{13.148}
$$

13.8 Local Optimization with Respect to Complex Parameters

In a variety of scenarios, we need to find the local optimal with respect to some complex parameter of a real metric that is a function. For real parameters, we would simply evaluate the derivative of the metric with respect to the parameter and find the parametric value whose derivative is zero. For complex parameters, this is a more complicated discussion. First, given our typical understanding of derivatives (requiring that the derivative is independent of the approach to evaluation point), a special requirement is placed on the function: The function must be analytic. Typically, the metrics in which we are interested are not analytic.

We do have an alternative approach. Imagine that our real metric is a function of our complex parameter a. For our metric to be real, it must also be a function of the complex conjugate of the parameter. To be explicit, we denote the metric function $f(a, a^*)$. Because we can only safely evaluate the derivatives with respect to real parameters, we can say that our parameter is a function of some real parameter α, so $a = a(\alpha)$. We will never specify this functionality explicitly. We can then evaluate the derivative as if $a(\alpha)$ and $a^*(\alpha)$ are different functions. We can then formally evaluate the derivative of $f(a, a^*)$ with respect to α,

$$\frac{\partial}{\partial \alpha} f(a[\alpha], a^*[\alpha]) = \frac{\partial}{\partial a} f(a[\alpha], a^*[\alpha]) \frac{\partial}{\partial \alpha} a[\alpha] + \frac{\partial}{\partial a^*} f(a[\alpha], a^*[\alpha]) \frac{\partial}{\partial \alpha} a^*[\alpha]$$

$$= g(a[\alpha], a^*[\alpha]) \frac{\partial}{\partial \alpha} a[\alpha] + g^*(a[\alpha], a^*[\alpha]) \frac{\partial}{\partial \alpha} a^*[\alpha]. \quad (13.149)$$

If we find the zero of $g^*(a[\alpha], a^*[\alpha])$, then this is also the zero of $g(a[\alpha], a^*[\alpha])$ and for any bounded value of $\partial/\partial \alpha\, a[\alpha]$, and this will be the zero of the entire function. Consequently, the solution is often to find the zero of $g^*(a[\alpha], a^*[\alpha])$ and solve for a. As an example, let us consider the simple quadratic form

$$f(a[\alpha], a^*[\alpha]) = (a[\alpha] - 1)(a^*[\alpha] - 1)$$

$$0 = \frac{\partial}{\partial \alpha} f(a[\alpha], a^*[\alpha])$$

$$= \left(\frac{\partial}{\partial \alpha} a[\alpha]\right) (a^*[\alpha] - 1) + (a[\alpha] - 1) \left(\frac{\partial}{\partial \alpha} a^*[\alpha]\right) \quad (13.150)$$

which is solved by

$$g^*(a[\alpha], a^*[\alpha]) = (a[\alpha] - 1) = 0 \quad (13.151)$$

$$a = 1. \quad (13.152)$$

We can extend this approach to the function of complex vectors of parameters $\mathbf{a}$ by considering the metric $f(\mathbf{a}[\alpha], \mathbf{a}^H[\alpha])$.

13.9   Volume Integrals

The integral over a volume (in this case, area) in the complex plane can be formally denoted by

$$\int d\Re\{z\}\, d\Im\{z\} f(z). \quad (13.153)$$

As an example, the probability density function for a circularly symmetric complex Gaussian distribution (discussed further in Section 14.1.6) in terms of $x = \Re\{z\}$ and $y = \Im\{z\}$ is given by

$$p(x, y)\, dx\, dy = \frac{1}{\pi \sigma^2} e^{-(x^2 + y^2)/\sigma^2}\, dx\, dy. \quad (13.154)$$

This equation can be constructed as the product of two independent, real Gaussian distributions by noting that the variance of the magnitude of the complex distribution

$\sigma^2 = \sigma_{Re}^2 + \sigma_{Im}^2$ (where σ_{Re}^2 is variance of the real part of z and σ_{Im}^2 is variance of the imaginary part of z) is twice the variance of either of the two real distributions σ_{Re} so that $\sigma^2 = 2\sigma_{Re}$. Consequently, the probability density is given by

$$p(x,y)\,dx\,dy = \frac{1}{\sqrt{\pi\sigma_{Re}^2}}e^{-x^2/(2\sigma_{Re}^2)}\frac{1}{\sqrt{\pi\sigma_{Re}^2}}e^{-y^2/(2\sigma_{Re}^2)}\,dx\,dy. \tag{13.155}$$

As a result, the integral over x and y is given by

$$\int dx\,dy\,p(x,y) = \int dx\,dy\,\frac{1}{\sqrt{2\pi\sigma_{Re}^2}}e^{-x^2/(2\sigma_{Re}^2)}\frac{1}{\sqrt{2\pi\sigma_{Re}^2}}e^{-y^2/(2\sigma_{Re}^2)}$$

$$= 1\cdot 1 = 1. \tag{13.156}$$

13.10 Special Functions

In this section, we briefly summarize some special functions that are often encountered in wireless communications in general and in this text in particular.

13.10.1 Sinc Function

The sinc is an important function that results from the Fourier transform of a top hat function. This function appears remarkably often in communications theory. There are two common definitions that differ by a scaling of π of the argument. We use the definition that is common in the engineering literature, given by

$$\text{sinc}(x) = \frac{\sin(\pi x)}{\pi x}. \tag{13.157}$$

13.10.2 Gamma Function

The gamma function $\Gamma(z)$ is an extension of the factorial function for real and complex numbers and is defined as

$$\Gamma(z) = \int_0^\infty d\tau\,\tau^{z-1}e^{-\tau}. \tag{13.158}$$

The integral requires analytic continuation to evaluate in the left half plane, and the gamma function is not defined for non-positive integer real values of z. For the special case of integer arguments, the gamma function can be expressed in terms of the factorial:

$$\Gamma(n) = (n-1)! \tag{13.159}$$

$$= \prod_{m=1}^{n-1} m. \tag{13.160}$$

While the values for the gamma function are unremarkable for non-negative integers, its value in the complex plane is interesting and connects to numerous topics in mathematics that are beyond the scope of this discussion.

Two functions that are related to the gamma function are the upper and lower incomplete gamma functions, which are defined respectively as

$$\Gamma(s, x) = \int_x^\infty d\tau \, \tau^{s-1} e^{-\tau} \tag{13.161}$$

and

$$\gamma(s, x) = \int_0^x d\tau \, \tau^{s-1} e^{-\tau} \,. \tag{13.162}$$

The lower incomplete gamma function $\gamma(s, x)$ is often encountered in communications systems in the cumulative distribution function of a χ^2 distributed random variable, which is proportional to the lower incomplete gamma function. The following asymptotic expansion of the lower incomplete gamma function is useful for analyzing the probability of error of wireless communications systems.

$$\gamma(s, x) = \frac{1}{s} x^s + o(x^s) \,. \tag{13.163}$$

13.10.3 Bessel Functions

Bessel functions [33–35] are given by the functional solutions for $f_\alpha(x)$ to the differential equation

$$x^2 \frac{\partial^2}{\partial x^2} f_\alpha(x) + x \frac{\partial}{\partial x} f_\alpha(x) + (x^2 - \alpha^2) f_\alpha(x) = 0 \,. \tag{13.164}$$

There are two "kinds" of solutions to this equation. Solutions of the first kind are denoted by $J_\alpha(x)$, and solutions of the second kind are denoted by $Y_\alpha(x)$ (and sometimes by $N_\alpha(x)$). The parameter α indicates the "order" of the function. The Bessel function of the second kind is defined in terms of the first kind by

$$Y_\alpha(x) = \frac{J_\alpha(x) \cos(\alpha \pi) - J_{-\alpha}(x)}{\sin(\alpha \pi)} \,. \tag{13.165}$$

Bessel functions are often the result of integrals of exponentials of trigonometric functions. As an example, we depict the first kind of Bessel function in Figure 13.5. Interestingly, the zeroth-order Bessel function of the first kind is related to a sinc function so that $J_0(x) = \text{sinc}(x/\pi)$.

Modified Bessel functions of the first and second kinds are denoted by $I_\alpha(x)$ and $K_\alpha(x)$, respectively. The modified Bessel function of the first kind is proportional to the Bessel function with a transformation in the complex plane of the form

$$I_\alpha(x) = (i)^{-\alpha} J_\alpha(ix) \,. \tag{13.166}$$

The modified Bessel function of the second kind is given by

$$K_\alpha(x) = \frac{\pi}{2} (i)^{\alpha+1} \left[J_\alpha(ix) + i Y_\alpha(ix) \right] \,. \tag{13.167}$$

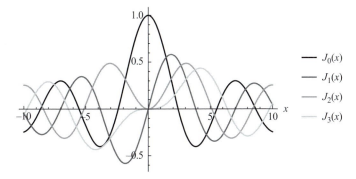

Figure 13.5 Bessel functions of the first kind.

As an example, we depict the modified Bessel function of the first kind in Figure 13.6.

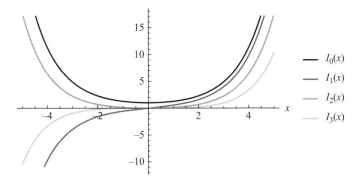

Figure 13.6 Modified Bessel function of the first kind.

13.10.4 Error Function

We denote the error function erf($\cdot$). It is defined by

$$\text{erf}(x_0) = \frac{2}{\sqrt{\pi}} \int_0^{x_0} dx\, e^{-x^2}. \tag{13.168}$$

13.10.5 Gaussian Q-Function

We denote the Gaussian Q-function by $Q(\cdot)$. The function integrates the area under a zero-mean unit-variance Gaussian distribution (discussed further in Section 14.1.6) from some value x_0 to infinity. It is defined by

$$Q(x_0) = \frac{1}{\sqrt{2\pi}} \int_{x_0}^{\infty} dx\, e^{-\frac{x^2}{2}} \tag{13.169}$$

$$= \frac{1}{2}\left[1 - \text{erf}\left(\frac{x_0}{\sqrt{2}}\right)\right]. \tag{13.170}$$

We display the erf(x) and Gaussian Q(x) functions in Figure 13.7.

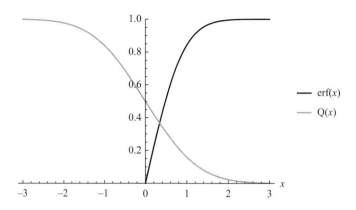

Figure 13.7 erf(x) and Gaussian Q(x) functions.

13.10.6   Marcum Q-Function

We denote the generalized Marcum Q-function by $Q_M(v, \mu)$ [10, 36–38]. It is defined by

$$Q_M(v, \mu) = \frac{1}{v^{M-1}} \int_\mu^\infty dy\, y^M\, e^{-\frac{x^2+v^2}{2}} I_{M-1}(v\, y)$$

$$= e^{-\frac{x^2+v^2}{2}} \sum_{m=1-M}^\infty \left(\frac{v}{\mu}\right)^m I_m(\mu\, v), \tag{13.171}$$

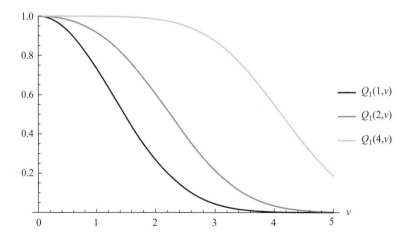

Figure 13.8 Marcum Q-function $Q_1(v, \mu)$ for values of $v = 1, 2$, and 4.

where $I_m(x)$ is the modified Bessel function that we define in Section 13.10.3. If M is not specified, a value of 1 is assumed. We display examples of the Marcum Q-function in Figure 13.8.

Problems

13.1 For some matrix $\mathbf{M}$ with complex components a, b, c, and d,

$$\mathbf{M} = \begin{pmatrix} a & b \\ c & d \end{pmatrix}.$$

Evaluate
(a) the real part of $\mathbf{M}$
(b) $\mathbf{M}^T$
(c) $\mathbf{M}^*$
(d) $\mathbf{M}^H$
(e) $\mathbf{M}^{-1}$

13.2 Evaluate the Taylor series expansion of x about 0 for $\log_e(1+x)$.

13.3 Evaluate $(\text{diag}\{1, 2, 3, 4\})^{-1}$.

13.4 Evaluate

$$\text{vec} \begin{pmatrix} a & b \\ c & d \end{pmatrix}.$$

13.5 By assuming $\mathbf{A}$, $\mathbf{B}$, $\mathbf{C}$, and $\mathbf{D}$ are invertible, evaluate $(\mathbf{A}\,\mathbf{B}\,\mathbf{C}\,\mathbf{D})^{-1}$.

13.6 By assuming the matrix $\mathbf{A}$ is invertible, $\mathbf{b}$ is a column vector, and $c \neq 0$, evaluate

$$\begin{pmatrix} \mathbf{A} & \mathbf{b} \\ \mathbf{b}^H & c \end{pmatrix}^{-1}$$

so that in a partitioned matrix, elements are represented by only multiplies and inversions of scalars and $\mathbf{A}$. You will need to use the inversion of the partitioned matrix inverse formula.

13.7 Given column vectors $\mathbf{a}$ and $\mathbf{b}$ of the same dimension, evaluate

$$\text{tr}\left\{(\mathbf{a} \quad \mathbf{b})(\mathbf{a} \quad \mathbf{b})^H\right\}.$$

13.8 For $\mathbf{A} = \text{diag}\{1, 2, 3, 4\}$ and

$$\mathbf{B} = \begin{pmatrix} 1 & 0 & 0 & 0 \\ 0 & 1 & 0 & 0 \\ 0 & 0 & 4 & 3 \\ 0 & 0 & 2 & 1 \end{pmatrix},$$

evaluate $|\mathbf{A}\,\mathbf{B}|$ by separately evaluating the determinants of $\mathbf{A}$ and $\mathbf{B}$.

13.9 Given

$$
\mathbf{A} = \begin{pmatrix} 1 & 2 & 3 & 4 \\ 3 & 1 & 3 & 4 \\ 4 & 1 & 1 & 2 \\ 2 & 3 & 2 & 1 \end{pmatrix},
$$

evaluate $\mathbf{A} \odot \mathbf{I}$.

13.10 Evaluate

$$
\frac{\partial}{\partial x} \begin{pmatrix} x^3 & 2yx \\ \sin(x) & x^2 \end{pmatrix}.
$$

13.11 Evaluate the eigenvalues and eigenvectors for the identity matrix.

13.12 Evaluate the eigenvalues for the matrix $\mathbf{M}$ and $\mathbf{a} \in \mathbb{C}^{4 \times 1}$ given by

$$
\mathbf{M} = \mathbf{I} + \mathbf{a}\mathbf{a}^H.
$$

13.13 For the column vectors $\mathbf{a}_1$ and $\mathbf{a}_2$, the matrix $\mathbf{A}$ is given by

$$
\mathbf{A} = (\mathbf{a}_1 \quad \mathbf{a}_2).
$$

Given this definition of $\mathbf{A}$,

(a) construct the projection matrix $\mathbf{P}_{\mathbf{A}}^{\perp}$ that is orthogonal to the column space of $\mathbf{A}$; and

(b) for column vector $\mathbf{b}$ of the same dimension of $\mathbf{a}$, evaluate $\mathbf{P}_{\mathbf{A}}^{\perp} \mathbf{b}$.

13.14 Given complex parameter a, $c \in \mathbb{R} > 0$, and the function $f(a) = 1 - c\,\|a - 1 - i\|^2$, use real parameter derivatives to evaluate the value of a that maximizes the function.

13.15 Evaluate the integral for $\Gamma(3)$ and show that it is given by $(3 - 1)!$.

14   Probability and Statistics

In this chapter we review probability and statistics. We define the ideas of probability, the probability density function, and the cumulative density function. We introduce Bayes' theorem that relates prior, marginal, posterior, and conditional probability densities, which allows us to define the maximum a posterior and the maximum likelihood estimators. We discuss central moments of distributions. We review multivariate probability distributions including the correlated multivariate complex circularly symmetric Gaussian distributions. We also review a set of useful distributions, including Rayleigh, exponential, χ^2, and Rician. Finally, we discuss random processes.

14.1   Probability

We think of probability as a way of characterizing the behavior of random variables. As a simple example, if we roll an honest six-sided die, one face will be on top after a roll. Each time we roll, we get an independent draw from the set of possible faces. The probability of any given side is equally likely and shows up 1/6 of the time, on average.

For many of the problems that we consider, the random variable is continuous. We can consider the probability of the random variable falling within the region of x and $x + \Delta$. If we let Δ go to zero, unless there is a delta function in the region (which does happen in some scenarios), then the probability of falling within the region goes to zero. This value is not particularly informative, but if we divide the probability by the interval Δ, then we have a probability density, which gives a distribution of probabilities for various intervals of potential values. If we evaluate this for all allowed values of x, then we have evaluated the probability density function (PDF), which provides a deterministic characterization of the random variable. We sometimes denote the PDF as $p_X(x)$ for the random variable X with a value near x. For some very small value of $\Delta \to dx$, the probability of falling into the region x and $x + dx$ is given by $p_X(x)\,dx$. Our random variable is described by the PDF, but any given draw from that distribution is just a number. As a simple example, the uniform probability distribution has a constant probability density over some region. If the probability for the random variable X for any value x is constant from the values of 1 to 1.5 and zero everywhere else, then the PDF is given by

$$p_X(x) = \begin{cases} 2 & ; & 1 \le x \le 1.5 \\ 0 & ; & \text{otherwise.} \end{cases} \tag{14.1}$$

For the sake of introduction, it is helpful to differentiate between a particular draw from a distribution given by x and the underlying random variable X. Later, we will often assume that the distinction is clear from the context and not be as pedantic. We indicate the probability of a set of outcomes with the notation $\Pr\{\cdot\}$. The random variable contains all the information required to understand the probability $\Pr\{x \in \mathcal{S}; \boldsymbol{\theta}\}$ that a given value x is in some set $\mathcal{S}$ of possible values given the parameters $\boldsymbol{\theta}$. Similarly, X is associated with the PDF $p_X(x; \boldsymbol{\theta})$ for some set of parameters. Consequently, the probability that the draw x from the random variable X is within the set of values $\mathcal{S}$ is given by

$$\Pr\{x \in \mathcal{S}; \boldsymbol{\theta}\} = \int_{\mathcal{S}} dx\, p_X(x; \boldsymbol{\theta}). \tag{14.2}$$

Depending upon the situation, the explicit dependency upon parameters may be suppressed. The cumulative distribution function (CDF) is the probability $P_X(x_0)$ that some random variable X is less than some threshold x_0:

$$P_X(x_0) = \Pr\{x \le x_0\}$$
$$= \int_{-\infty}^{x_0} dx\, p_X(x; \boldsymbol{\theta}). \tag{14.3}$$

The probability of something occurring is bounded by 0 and 1. For any value of x, the PDF is greater than or equal to 0. However, at any given value, the PDF can be arbitrarily large so long as the integral over some set of values is bounded by 1. These large values can lead to confusion, so it may be helpful to remember that the PDF does not have units of probability. Rather, its units are probability per units of X.

14.1.1   Bayes' Theorem

When random variables x and y are related and dependent, we can explicitly explain their relationships using Bayes' theorem. To do this decomposition, we need to define a few different types of functions for probability density. These densities include: prior probability density,[1] marginal probability density, posterior probability density,[2] and conditional probability density, which are denoted here as

$$p_X(x): \text{ prior probability density}$$
$$p_Y(y): \text{ marginal probability density}$$
$$p_X(x|y): \text{ posterior probability density}$$
$$p_Y(y|x): \text{ conditional probability density.} \tag{14.4}$$

The prior probability density $p_X(x)$ is the probability for X independent of the value of Y. The marginal probability density $p_Y(y)$ is the probability for Y independent

---

[1] Often the Latin form is used, *a priori*, to denote the probability.
[2] Often the Latin form is used, *a posteriori*, to denote the probability.

of the value of X. The posterior probability density $p_X(x|y)$ is the probability for X under the assumption that y has a particular given value. The conditional probability density $p_Y(y|x)$ is the probability for Y under the assumption that x has a particular given value.

For single variables, this relationship (Bayes' theorem) can be written in the important form

$$p_X(x|y) = \frac{p_Y(y|x) p_X(x)}{p_Y(y)}. \tag{14.5}$$

A useful interpretation of this form is to consider random variable X the input to a random variable that produces random variable Y. Thus, the likelihood of a given value x for the random input variable X is found given observation y of the output distribution Y. Throughout statistical signal processing research, a common source of debate is whether to use a Bayesian approach that requires knowledge of the prior distribution or to use a suboptimal, but potentially more robust, form. Mismatch between the assumed and real priors can affect the performance of various algorithms.

14.1.2 MAP and ML Estimators

Bayes' theorem is particularly useful because, for our applications, we want to estimate a parameter or information at the output of some process. To do this, it makes sense to find the input variable estimate of x, denoted as $\hat{x}$ given some observation y, by maximizing the posterior probability density $p_X(x|y)$. Thus,

$$\hat{x} = \text{argmax}_x \, p_X(x|y), \tag{14.6}$$

where $\text{argmax}_x \, f(x)$ provides the value x that maximizes some function $f(x)$. The result in Equation (14.6) is the maximum a posteriori (MAP) estimate. However, because we have limited information about the prior or marginal probabilities, we often optimize the conditional probability,

$$\hat{x} = \text{argmax}_x \, p_Y(y|x), \tag{14.7}$$

which is called the maximum likelihood (ML) estimate. The approach seems rather strange because, in some sense, we are optimizing the conditional probability by flipping the inputs and outputs as we usually think about them, but we see in Bayes' theorem in Equation (14.5) that this approach is reasonable because the posterior density grows with the conditional density.

14.1.3 Change of Variables

For some applications, we have a random variable that has known characteristics, but we want to evaluate the characteristics of a new random variable that is functionally related to the original. We consider a random variable X with probability density $p_X(x)$ and a new random variable that is a function of $Y = f(x)$. By assuming that the function $f(x)$ is one to one and differentiable, we can find the probability density $p_Y(y)$ of Y using the following transformation:

$$p_Y(y) = \frac{p_X(f^{-1}(y))}{\left\| \frac{\partial}{\partial x} f(x) \big|_{x=f^{-1}(y)} \right\|},$$ (14.8)

where $x = f^{-1}(y)$ indicates the inverse function of $f(x)$ and the notation $\cdot|_{x=x_0}$ indicates evaluating the expression to the left with the value x_0. However, it is not uncommon for the inverse to have multiple solutions. If the jth solution given by x at some value y to the inverse is given by $x = f_j^{-1}(y)$, then the transformation of densities is given by

$$p_Y(y) = \sum_j \frac{p_X(f_j^{-1}(y))}{\left\| \frac{\partial}{\partial x} f(x) \big|_{x=f_j^{-1}(y)} \right\|}.$$ (14.9)

As an example let us consider the real, unit-variance, zero-mean Gaussian distribution for X with a change of variable to Y such that $y = x^2$. Here, we have to keep in mind that both positive and negative values of x map to the same value of y. Consequently, we use the form given in Equation (14.9). We define the functions

$$y = f(x) = x^2$$ (14.10)

$$\frac{\partial}{\partial x} f(x) = 2x$$ (14.11)

$$f_+^{-1}(y) = \sqrt{y}$$ (14.12)

$$f_-^{-1}(y) = -\sqrt{y}.$$ (14.13)

By starting with our Gaussian distribution, we can convert this to a distribution for Y, given by

$$p_X(x) = \frac{1}{\sqrt{2\pi}} e^{-x^2/2}$$ (14.14)

$$p_Y(y) = \sum_{\pm} \frac{p_X(f_j^{-1}(y))}{\left\| \frac{\partial}{\partial x} f(x) \big|_{x=f_{\pm}^{-1}(y)} \right\|}$$

$$= \frac{1}{\sqrt{2\pi}} \left(\frac{e^{-(f_+^{-1}(y))^2/2}}{\left\| \frac{\partial}{\partial x} f(x) \big|_{x=f_+^{-1}(y)} \right\|} + \frac{e^{-(f_-^{-1}(y))^2/2}}{\left\| \frac{\partial}{\partial x} f(x) \big|_{x=f_-^{-1}(y)} \right\|} \right)$$

$$= \frac{1}{\sqrt{2\pi}} \left(\frac{e^{-(f_+^{-1}(y))^2/2}}{\left\| 2x \big|_{x=f_+^{-1}(y)} \right\|} + \frac{e^{-(f_-^{-1}(y))^2/2}}{\left\| 2x \big|_{x=f_-^{-1}(y)} \right\|} \right)$$

$$= \frac{1}{\sqrt{2\pi}} \left(\frac{e^{-y/2}}{\|2\sqrt{y}\|} + \frac{e^{-y/2}}{\|-2\sqrt{y}\|} \right) = \frac{1}{\sqrt{2\pi}} \frac{e^{-y/2}}{\sqrt{y}} \quad ; y \geq 0.$$ (14.15)

It is worth noting that the PDF for our new random variable Y is known as a χ^2 distribution with a single degree of freedom, as discussed in Section 14.1.9.

We can extend this change of variable idea to include multivariate distributions. To make the notation easier, we often denote a group of random variables with random vectors or matrices. We consider a multivariate distribution involving M random variables, which is discussed in greater detail in Section 14.1.5. We define the random vectors $\mathcal{X} \in \mathbb{R}^{M \times 1}$, whose realizations are denoted by $\mathbf{x}$, and $\mathcal{Y} \in \mathbb{R}^{M \times 1}$ whose realizations are denoted by $\mathbf{y}$. The PDFs of these random vectors are given by $p_{\mathcal{X}}(\mathbf{x})$ and $p_{\mathcal{Y}}(\mathbf{y})$, respectively. We define a vector function $\mathbf{f}(\mathbf{x})$ that maps $\mathbf{x}$ to $\mathbf{y}$ such that $\mathbf{y} = \mathbf{f}(\mathbf{x})$. If, for a value $\mathbf{y}$, there are multiple solutions for $\mathbf{x}$, then the functional inverse $\mathbf{f}_j^{-1}(\mathbf{y})$ is the jth solution. The relationship between PDFs of $\mathcal{X}$ and $\mathcal{Y}$ is then given by

$$p_{\mathcal{Y}}(\mathbf{y}) = \sum_j \frac{p_{\mathcal{X}}(\mathbf{f}_j^{-1}(\mathbf{y}))}{\left\| \left| \frac{\partial \mathbf{f}(\mathbf{x})}{\partial \mathbf{x}} \right|_{\mathbf{x} = \mathbf{f}_j^{-1}(\mathbf{y})} \right\|}, \tag{14.16}$$

where $|\partial \mathbf{f}(\mathbf{x})/\partial \mathbf{x}|$ indicates the determinant of the Jacobian matrix associated with the two random vectors. The notation $\| \cdot \|_{x=x_0}$ indicates the absolute value of the quantity within the bars evaluated at $x = x_0$. The Jacobian matrix notation $\partial \mathbf{f}(\mathbf{x})/\partial \mathbf{x}$ is an efficient notation for

$$\frac{\partial \mathbf{f}(\mathbf{x})}{\partial \mathbf{x}} = \begin{pmatrix} \frac{\partial \{\mathbf{f}(\mathbf{x})\}_1}{\partial x_1} & \frac{\partial \{\mathbf{f}(\mathbf{x})\}_1}{\partial x_2} & \cdots & \frac{\partial \{\mathbf{f}(\mathbf{x})\}_1}{\partial x_M} \\ \frac{\partial \{\mathbf{f}(\mathbf{x})\}_2}{\partial x_1} & \frac{\partial \{\mathbf{f}(\mathbf{x})\}_2}{\partial x_2} & \cdots & \frac{\partial \{\mathbf{f}(\mathbf{x})\}_2}{\partial x_M} \\ & & \vdots & \\ \frac{\partial \{\mathbf{f}(\mathbf{x})\}_M}{\partial x_1} & \frac{\partial \{\mathbf{f}(\mathbf{x})\}_M}{\partial x_2} & \cdots & \frac{\partial \{\mathbf{f}(\mathbf{x})\}_M}{\partial x_M} \end{pmatrix}, \tag{14.17}$$

where $x_m = \{\mathbf{x}\}_m$ indicates the random variable associated with the mth entry in $\mathbf{x}$.

14.1.4   Central Moments of a Distribution

In some situations, we only need an approximate characterization of a random variable. In other situations, we can completely characterize our random variable with a small number of parameters. For example, we can completely describe a Gaussian random variable (discussed further in Section 14.1.6) with its mean and variance. We often describe the characteristics of a distribution of random variables by using a subset of moments about the mean of the distribution. We indicate the expectation of some function $f(x)$ of the random variable X with the PDF $p_X(x)$ by

$$E[f(X)] = \int dx f(x) p_X(x). \tag{14.18}$$

The mean value of the random variable X is given by

$$E[X] = \int dx \, x \, p_X(x). \tag{14.19}$$

We indicate the mth central moment about the mean by μ_m, given by

$$\mu_m = \int dx \, (x - E[X])^m \, p_X(x). \tag{14.20}$$

We use the word *central* to indicate that fluctuations are measured with respect to the mean, which is why we use the form $(x - E[X])$.

By construction, the value of μ_1 is zero. The following central moments, denoted here as μ_2, μ_3, and μ_4, are related to the variance, skew, and kurtosis excess of a distribution.

The variance of random variable X is given by

$$\sigma_X^2 = \int dx\,(x - E[X])^2\,p_X(x)$$

$$= \mu_2\,.\tag{14.21}$$

Note that, in situations where the random variable of concern is clear, we shall omit the subscript, denoting the variance simply as σ^2. The variance, as the name implies, provides the most basic description of how much a signal varies about its mean. For many of the random variables that we see in our applications, the mean is zero. Consequently, we observe that the power of a signal is proportional to its amplitude variance.

We use the skewness of random variable X to indicate the asymmetry of a distribution about its mean. The skewness is given by the third central moment normalized by the variance to the 3/2 power, and is thus unitless:

$$\text{skew}\{X\} = \frac{\int dx\,(x - E[X])^3\,p_X(x)}{\left(\sigma^2\right)^{3/2}}$$

$$= \frac{\mu_3}{\left(\sigma^2\right)^{3/2}}\,.\tag{14.22}$$

If the distribution has a large extent to the right (larger values) of its mean but a small extent to its left (smaller values) of its mean, then it has a positive skew. Conversely, if the distribution has a larger extent on the left, then it has a negative skew. For example, the zero-mean unit-variance triangle distribution, which we display in Figure 14.1, is given by

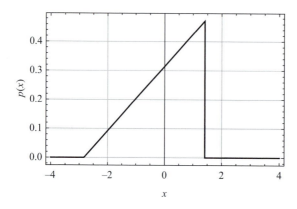

Figure 14.1 Skew of a zero-mean unit-variance triangle distribution that has a skew of $-2\sqrt{2}/5$.

$$p(x) = \begin{cases} \frac{2\sqrt{2}+x}{9} & ; \quad \|1 + \sqrt{2}x\| < 3 \\ 0 & ; \quad \text{otherwise} \end{cases} \tag{14.23}$$

and has a skew of $-2\sqrt{2}/5$.

Finally, the kurtosis of random variable X is a measure of a distribution's "peakiness." It is given by the fourth central moment normalized by the variance squared; thus, it is unitless. The excess kurtosis is given by subtracting 3 from the kurtosis:

$$\text{excess kurtosis}\{X\} = \frac{\int dx\,(x - E[X])^4\,p_X(x)}{\left(\sigma^2\right)^2} - 3$$

$$= \frac{\mu_4}{\left(\sigma^2\right)^2} - 3, \tag{14.24}$$

which has the desirable characteristic of evaluating to zero for Gaussian distributions (discussed further in Section 14.1.6). Unfortunately, there is sometimes confusion in the literature as to whether *kurtosis* indicates kurtosis or excess kurtosis. Given our definition of excess kurtosis, such that a Gaussian distribution has an excess kurtosis of zero, distributions with positive kurtosis (leptokurtic) are more "peaky" than a Gaussian distribution with the same variance, and distributions with negative kurtosis (platykurtic) have "broader shoulders" and a lower peak than a Gaussian distribution with the same variance. We display examples in Figure 14.2 for Gaussian (with excess kurtosis of 0), double-sided exponential (with excess kurtosis of 3), and top hat (with excess kurtosis of $-6/5$) distributions, which are given by

$$p_G(x) = \frac{1}{\sqrt{2\pi}} e^{-x^2/2}, \tag{14.25}$$

$$p_{DE}(x) = \frac{\sqrt{2}}{2} e^{-\|x\|\sqrt{2}}, \tag{14.26}$$

and

$$p_{TH}(x) = \begin{cases} \frac{1}{2\sqrt{3}} & ; \quad \|x\| < \sqrt{3} \\ 0 & ; \quad \text{otherwise} \end{cases} \tag{14.27}$$

respectively.

14.1.5   Multivariate Probability Distributions

The probability density function of multiple random variables indicates the probability that values of the random variables are within some infinitesimal hypervolume about some point in the variable space. The probability density is denoted as

$$p_{X_1, X_2, \cdots}(x_1, x_2, \cdots). \tag{14.28}$$

If the random variables are independent, then the joint probability density function is equal to the product of the individual probability densities:

$$p_{X_1, X_2, \cdots}(x_1, x_2, \cdots) = \prod_m p_{X_m}(x_m). \tag{14.29}$$

In many cases, the sequence of random variables is drawn independently from the same (or identical) distribution. We say that these are independent and identically

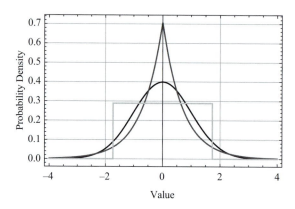

Figure 14.2 Comparison of excess kurtoses of real random variables drawn from the PDFs: Gaussian (black) with excess kurtosis of 0, double-sided exponential (dark gray) with excess kurtosis of 3, and top hat (light gray) with excess kurtosis of $-6/5$. The examples are zero-mean unit-variance distributions.

distributed (i.i.d.) random variables. Given some parameter $\mathbf{A}$, the probability of a complex random matrix variable $\mathcal{X}$ having a value $\mathbf{X}$ that is contained within a space defined by $\mathcal{S}$ is given by

$$P_{\mathcal{X}}(\mathbf{X} \in \mathcal{S}; \mathbf{A}) = \int_{\mathcal{S}} d\Omega_{\mathbf{X}}\, p_{\mathcal{X}}(\mathbf{X}; \mathbf{A}),$$

$$= \int_{\mathcal{S}} d\Omega_{\mathbf{X}}\, p_{\mathcal{X}}((\mathbf{X})_{1,1}, (\mathbf{X})_{1,2}, \cdots (\mathbf{X})_{2,1}, \cdots ; \mathbf{A}), \qquad (14.30)$$

where we use the notation $d\Omega_{\mathbf{X}}$ for the differential hypervolume built from the product of all the differentials associated with all the random variables. For complex random matrix $\mathbf{X}$, it is given by

$$d\Omega_{\mathbf{X}} = d\{\Re\mathbf{X}\}_{1,1}\, d\{\Re\mathbf{X}\}_{1,2} \cdots d\{\Re\mathbf{X}\}_{m,n}$$

$$\cdot d\{\Im\mathbf{X}\}_{1,1}\, d\{\Im\mathbf{X}\}_{1,2} \cdots d\{\Im\mathbf{X}\}_{m,n}. \qquad (14.31)$$

Note that the measure is expressed in terms of the real and imaginary components of the complex random variable. This convention is not employed universally but typically will be assumed within this text. In the case of a real random variable, the imaginary differentials are dropped.

The probability density function of a given set of random variables x_m given or conditioned on particular values for another set of variables y_n is denoted by

$$p_{X_1, X_2, \cdots}(x_1, x_2, \cdots | y_1, y_2 \cdots). \qquad (14.32)$$

We relate the conditional and prior probability densities using

$$p_{X_1, X_2, \cdots, Y_1, Y_2, \cdots}(x_1, x_2, \cdots, y_1, y_2 \cdots)$$

$$= p_{X_1, X_2, \cdots}(x_1, x_2, \cdots | y_1, y_2 \cdots)\, p_{Y_1, Y_2, \cdots}(y_1, y_2 \cdots)$$

$$= p_{Y_1, Y_2, \cdots}(y_1, y_2 \cdots | x_1, x_2, \cdots)\, p_{X_1, X_2, \cdots}(x_1, x_2, \cdots). \qquad (14.33)$$

14.1.6 Gaussian Distribution

The Gaussian (or normal) distribution is an essential distribution in signal processing. Because of the central limit theorem, processes that combine the effects of many distributions often converge to the Gaussian distribution. Also, for a given mean and variance, the entropy (which is used in the evaluation of mutual information) is maximized for Gaussian distributed variables. We discussed this in greater detail in Section 3.2.1. The analysis of multiple-antenna systems will often take advantage of multivariate Gaussian distributions. The PDF for a real random variable X with value x, mean $\mu = E[X]$, and variance $\sigma^2 = E[(X - \mu)^2]$ is given by

$$p_X(x; \mu, \sigma)\, dx = \frac{1}{\sqrt{2\pi\sigma^2}} e^{-\frac{(x-\mu)^2}{2\sigma^2}}\, dx. \tag{14.34}$$

This normal (or Gaussian) distribution is often identified by $\mathcal{N}(\mu, \sigma^2)$. The complex normal (or Gaussian) distribution assuming circular symmetry for a complex random variable Z with value z, mean μ, and variance $\sigma^2 = E[\|z - \mu\|^2]$ is given by

$$p_Z(z; \mu, \sigma)\, d\Re z\, d\Im z = \frac{1}{\pi\sigma^2} e^{-\frac{\|z-\mu\|^2}{\sigma^2}}\, d\Re z\, d\Im z. \tag{14.35}$$

The complex version of the distribution is often denoted by $\mathcal{CN}(\mu, \sigma^2)$. As an example, we depict in Figure 14.3 the PDF for a circularly symmetric complex Gaussian distribution with a mean of $\mu = 2 + i/2$ and variance of $\sigma^2 = 4$.

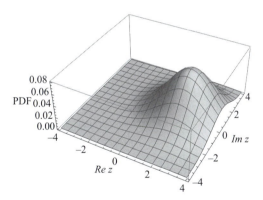

Figure 14.3 Graphical representation of the PDF of a circularly symmetric complex Gaussian distribution with a mean of $\mu = 2 + i/2$ and variance of $\sigma^2 = 4$.

The probability density for an m by k random matrix $\mathcal{Z}$ with value $\mathbf{Z} \in \mathbb{C}^{m \times k}$ drawn from a multivariate complex Gaussian is given by

$$p_{\mathcal{Z}}(\mathbf{Z}; \mathbf{X}, \mathbf{R})\, d\Omega_{\mathbf{Z}} = \frac{1}{|\mathbf{R}|^k\, \pi^{mk}} e^{-\mathrm{tr}\{(\mathbf{Z}-\mathbf{X})^H\, \mathbf{R}^{-1}\, (\mathbf{Z}-\mathbf{X})\}}\, d\Omega_{\mathbf{Z}}, \tag{14.36}$$

where the mean of $\mathcal{Z}$ is given by $\mathbf{X} \in \mathbb{C}^{m \times k}$. The covariance of the rows of the random matrix is $\mathbf{R} \in \mathbb{C}^{m \times m}$ under the assumption of independent columns (note that it is possible to define a more general form of Gaussian random matrix with dependent columns). The notation $d\Omega_{\mathbf{Z}}$ indicates the differential hypervolume in terms of the real parameters $\Re\{\mathbf{Z}\}_{p,q}$ and $\Im\{\mathbf{Z}\}_{p,q}$, indicating the real and imaginary part of the elements of $\mathbf{Z}$. Here, p and q are indices identifying elements in the matrix. As a reminder, $|\mathbf{R}|$ indicates the determinant of $\mathbf{R}$.

The covariance matrix is an important concept used repeatedly throughout adaptive communications. For some complex random matrix $\mathcal{Z}$ with values $\mathbf{Z}$ and mean $\mathbf{X}$, the covariance matrix is given by

$$\mathbf{R} = E\left[\frac{(\mathcal{Z} - E[\mathcal{Z}])(\mathcal{Z} - E[\mathcal{Z}])^H}{k}\right]$$

$$= \int d\Omega_{\mathbf{Z}} \frac{(\mathbf{Z} - \mathbf{X})(\mathbf{Z} - \mathbf{X})^H}{k} p_{\mathcal{Z}}(\mathbf{Z}; \mathbf{X}). \tag{14.37}$$

The covariance is Hermitian, $\mathbf{R} = \mathbf{R}^H$. In this form, the covariance matrix is a measure of the cross-correlation between the elements along the columns of $\mathbf{Z}$.

14.1.7   Rayleigh Distribution

The magnitude of a random variable drawn from a complex, circularly symmetric Gaussian distribution with variance σ^2 follows the Rayleigh distribution. Here, we will identify this random variable as Q with value q. Suppose the Gaussian variable is denoted by Z with value z and has variance σ^2. If its magnitude is denoted by

$$q = \|z\|, \tag{14.38}$$

then probability density[3] for the real Rayleigh variable Q is given by

$$p_{Ray}(q)\, dq = \begin{cases} \frac{2q}{\sigma^2} e^{-q^2/\sigma^2}\, dq & ; q \geq 0 \\ 0\, dq & ; \text{ otherwise,} \end{cases} \tag{14.39}$$

where σ^2 is the variance of the complex Gaussian variable z. The cumulative distribution function $P_{Ray}(q)$ for the Rayleigh variable q is given by

$$P_{Ray}(q) = \int_0^q dx\, p(x)$$

$$= \begin{cases} 1 - e^{-q^2/\sigma^2} & ; q \geq 0 \\ 0 & ; \text{ otherwise.} \end{cases} \tag{14.40}$$

14.1.8   Exponential Distribution

The square of a Rayleigh distributed random variable has an exponential distribution and is a useful quantity in analyzing the received power of a narrowband transmission through a Rayleigh fading channel because, in that case, the amplitude of the

---

[3] This density assumed that the complex variance is given by σ^2, which is different from a common assumption that the variance is given for a real variable. Consequently, there are some subtle scaling differences.

channel coefficient follows a Rayleigh distribution. The exponential random variable is parameterized by the inverse of its mean λ.

The PDF of the exponential random variable is

$$p_{Exp}(x)\, dx = \begin{cases} \lambda e^{-\lambda x}\, dx & ; \ x \geq 0 \\ 0\, dx & ; \ x < 0. \end{cases} \tag{14.41}$$

The cumulative distribution function of the exponential random variable is

$$P_{Exp}(x) = \begin{cases} 1 - e^{-\lambda x} & ; \ x \geq 0 \\ 0 & ; \ x < 0. \end{cases} \tag{14.42}$$

Its mean and variance are $\frac{1}{\lambda}$ and $\frac{1}{\lambda^2}$, respectively.

14.1.9 Central χ^2 Distribution

The sum q of the magnitude squared of real, zero-mean, unit-variance Gaussian variables X_m with value x_m is characterized by the χ^2 distribution. The sum of k independent Gaussian variables denoted by q is

$$q = \sum_{m=1}^{k} x_m^2. \tag{14.43}$$

Since the random variables X_m have unit variance,

$$E[\|X_m\|^2] = 1. \tag{14.44}$$

The distribution $p_{\chi^2}(q)$ for the sum of the magnitude square q of k independent, zero-mean, unit-variance real Gaussian random variables is given by

$$p_{\chi^2}(q; k)\, dq = \begin{cases} \frac{1}{2^{k/2}\,\Gamma(k/2)}\, q^{k/2-1}\, e^{-q/2}\, dq & ; \ q \geq 0 \\ 0\, dq & ; \ \text{otherwise.} \end{cases} \tag{14.45}$$

The cumulative distribution function $P_{\chi^2}(q; k)$ of the χ^2 random variable is given by

$$P_{\chi^2}(q; k) = \int_0^q dr\, p_{\chi^2}(r; k)$$
$$= \begin{cases} \frac{1}{\Gamma(k/2)}\, \gamma\!\left(\frac{k}{2}, \frac{q}{2}\right) & ; \ q \geq 0 \\ 0 & ; \ \text{otherwise,} \end{cases} \tag{14.46}$$

where $\Gamma(\cdot)$ is the standard gamma function and $\gamma(\cdot, \cdot)$ is the lower incomplete gamma function given by Equation (13.162).

14.1.10 Complex χ^2 Distribution

With a slight abuse of terminology, we define the complex χ^2 distribution as the distribution of the sum q of n independent complex, circularly symmetric Gaussian random variables Z_m with values z_m. The sum q is given by [39]:

$$q = \sum_{m=1}^{n} \|z_m\|^2 \tag{14.47}$$

$$E[\|Z_m\|^2] = \sigma^2 . \tag{14.48}$$

To be clear, the variance detailed here is in terms of the complex Gaussian variable and we include the variance explicitly as a parameter of the distribution. By employing Equation (14.8) and noting that the number of real degrees of freedom is twice the number of complex degrees of freedom ($k = 2n$), the distribution $p_{\chi^2}^{\mathbb{C}}(q; n, \sigma^2)$ for the sum of the magnitude squared $q \geq 0$ is given by

$$
\begin{aligned}
p_{\chi^2}^{\mathbb{C}}(q; n, \sigma^2) \, dq &= \frac{1}{\sigma^2/2} p_{\chi^2}\left(\frac{q}{\sigma^2/2}; 2n\right) dq \\
&= \frac{1}{\sigma^2 \, 2^{n-1} \, \Gamma(n)} \left(\frac{2q}{\sigma^2}\right)^{n-1} e^{-\frac{q}{\sigma^2}} \, dq \\
&= \frac{q^{n-1}}{\left(\sigma^2\right)^n \, \Gamma(n)} e^{-\frac{q}{\sigma^2}} \, dq ,
\end{aligned}
\tag{14.49}
$$

where it is assumed that the variance σ^2 of z_m is the same for all m and the density is zero for $q < 0$. The cumulative distribution for q is given by

$$
\begin{aligned}
P_{\chi^2}^{\mathbb{C}}(q; n, \sigma^2) &= \int_0^q dr \, p_{\chi^2}^{\mathbb{C}}(q; n, \sigma^2) \\
&= \frac{1}{\Gamma(n)} \gamma\left(n, \frac{q}{\sigma^2}\right) .
\end{aligned}
\tag{14.50}
$$

14.1.11   Rician Distribution

The magnitude of the sum of a real scalar a and a random complex variable z sampled from a circularly symmetric complex Gaussian distribution with zero mean and variance σ^2 is given by the random variable Y with value y:

$$
\begin{aligned}
y &= \|a + z\| \\
&= \sqrt{(a + \Re\{z\})^2 + \Im\{z\}^2} .
\end{aligned}
\tag{14.51}
$$

The random variable y follows the Rice or Rician distribution, whose PDF $p_{Rice}(y)$ is

$$
p_{Rice}(y) \, dy = \begin{cases} \frac{2y}{\sigma^2} I_0\left(\frac{2 a y}{\sigma^2}\right) e^{-(y^2 + a^2)/\sigma^2} \, dy & ; \, y \geq 0 \\ 0 \, dy & ; \, \text{otherwise} \end{cases}, \tag{14.52}
$$

where $I_0(\cdot)$ is the zeroth-order modified Bessel function of the first kind (discussed in Section 13.10.3). In channel phenomenology, it is common to describe this distribution in terms of the *Rician K-factor*, which is the ratio of the coherent to the fluctuation power,

$$K = \frac{a^2}{\sigma^2} . \tag{14.53}$$

It may be worth noting that the Rician distribution is often described in terms of two real Gaussian variables. Consequently, the distribution given here differs from the common form by replacing σ^2 with $\sigma^2/2$.

The cumulative distribution function for a Rician variable of value greater than zero is given by

$$
\begin{aligned}
P_{Rice}(y_0) &= \int_0^{y_0} dy \, p_Y(y) \\
&= \int_0^{y_0} dy \, \frac{2y}{\sigma^2} I_0 \left(\frac{2ay}{\sigma^2} \right) e^{-(y^2+a^2)/\sigma^2} \\
&= 1 - Q_{M=1} \left(\frac{a\sqrt{2}}{\sigma}, \frac{y_0\sqrt{2}}{\sigma} \right),
\end{aligned}
\tag{14.54}
$$

where $Q_M(v, \mu)$ is the Marcum Q-function discussed in Section 13.10.6. The distribution for the square of a Rician random variable $q = y^2$ is the complex noncentral χ^2 distribution with one complex degree of freedom.

14.2 Random Processes

While random variables are mappings from an underlying space of events to real numbers, a *random process* can be viewed as a mapping from an underlying space of events onto a sequence of values or functions. Random processes are essentially random functions and are useful to describe transmitted and received signals.

Consider a random process $X(t)$. A particular draw from the random process will produce a specific random curve $x(t)$. For any particular t, $X(t)$ is simply a random variable. A complete statistical characterization of a random process requires the description of the joint probability densities (or distributions) of the random variables $X(t)$ for all possible values of t. For most of our problems, we consider sampled signals, so we can apply the same discussion to a sequence or vector of random samples. We can label a given set of draws as $x_{t_1}, x_{t_2}, \cdots, x_{t_m}$.

14.2.1 Mean of Random Processes

The meaning of the mean of a random process is potentially ambiguous. We are used to thinking of a single random variable, so the mean is just the expected value of the random variable. For some random process $X(t)$, the mean is a function of time. We can think of averaging a given sequence or function or an ensemble of random universes. Thus, the mean $\mu(t)$ as a function of time t is given by

$$
\mu(t) = E[X(t)].
\tag{14.55}
$$

This is a particularly valuable concept for analyzing communications systems. If we consider the basic baseband flat-fading channel model, given by

$$
z(t) = a\, s(t) + n(t),
\tag{14.56}
$$

where $z(t)$ is the received signal as a function of time t, a is the complex attenuation, $s(t)$ is a deterministic transmitted signal, and $n(t)$ is the zero-mean noise, then the mean at a given time t_0 is given by

$$E[z(t_0)] = E[a\,s(t_0) + n(t_0)] = E[a\,s(t_0)] + E[n(t_0)]$$
$$= a\,s(t_0)\,. \tag{14.57}$$

This is a useful model for known transmitted sequences. However, for many problems the transmitted sequence is effectively unknown, and we consider $s(t)$ a random variable too. This would produce a different mean, so we need to understand our assumptions.

14.2.2　Stationary Random Processes

There are a number of special cases of random processes. We consider a random process to be stationary if the cumulative density function remains the same given some shift in time, so $P_X(t) = P_X(t - T)$. For a discretely sampled sequence we would say

$$P(x_{t_1}, x_{t_2}, \cdots, x_{t_m}) = P(x_{t_1 - T}, x_{t_2 - T}, \cdots, x_{t_m - T})\,. \tag{14.58}$$

This is a very strong requirement, and is often not valid for our problems. This leads us to a softer version identified as a wide-sense stationary random process that is sufficient and applicable for many problems of interest.

14.2.3　Wide-Sense Stationary Random Processes

Of particular interest are the second-order statistics of ergodic random processes. The autocorrelation function (which does not remove the mean in evaluation) is

$$R_X(\tau_1, \tau_2) = E\{X(t - \tau_1)\,X^*(t - \tau_2)\}\,. \tag{14.59}$$

It is also possible to define the cross-correlation between two random processes $X(t)$ and $Y(t)$ as follows

$$R_{XY}(\tau_1, \tau_2) = E\{X(\tau_1)\,Y^*(\tau_2)\}\,. \tag{14.60}$$

Note that the expectations above are taken with respect to the ensemble of possible realizations of the processes $X(t)$ and $Y(t)$, jointly.

Wide-sense stationary (WSS) random processes are random processes for which the mean function is a constant and the autocorrelation function (which does not remove the mean in its evaluation) is dependent only on the difference in the time indices. A random process $X(t)$ is WSS if the following two conditions hold so that the autocorrelation is only a function $f(\tau_2 - \tau_1)$ of the difference in time shifts that matter:

$$E[X(t)] = \mu_X \tag{14.61}$$
$$R_X(\tau_1, \tau_2) = f(\tau_2 - \tau_1)\,. \tag{14.62}$$

We exploit this model for the random process to evaluate the power spectral density in Sections 6.2.1, 15.4.13, and 15.6.2. We can extend this to jointly WSS for $X(t)$ and $Y(t)$ by requiring

$$E[X(t)] = \mu_X \tag{14.63}$$

$$E[Y(t)] = \mu_Y \tag{14.64}$$

$$R_{XY}(\tau_1, \tau_2) = f(\tau_2 - \tau_1). \tag{14.65}$$

Because all physical systems are dynamic, no system can be exactly represented by a stationary or WSS process; however, we often estimate system characteristics over a sufficiently short duration of time that stationary models are useful.

14.2.4   Ergodic Random Processes

In general, the joint density for all possible t is very difficult to obtain for real-world signals. If we restrict ourselves to *ergodic* processes that, loosely speaking, are random processes for which single realizations of the process contain the statistical properties of the entire ensemble, it is possible to estimate certain statistical properties of the ensemble from a single realization of the process.

Problems

14.1 Determine if the following functions $f(x)$ are valid PDFs and explain why:
(a) $f(x) = 1/(1 + x^2)$
(b) $f(x) = 1$ for $\|x\| < 1$ and $f(x) = 0$, otherwise
(c) $f(x) = \delta(x - 2)$
(d) $f(x) = 2/3 - x^2/2$ for $\|x\| < 1$ and $f(x) = 0$, otherwise
(e) $f(x) = \pi/4 \cos(\pi x) + \delta(x)/2$ for $\|x\| < 1/2$ and $f(x) = 0$, otherwise

14.2 Evaluate the marginal probability density $p_Y(y)$ if

$$p_Y(y|x) = \frac{1}{\sqrt{2\pi}} e^{-(y-x)^2/2},$$

and the prior probability density is given by

$$p_X(x) = \frac{1}{\sqrt{2\pi 3}} e^{-x^2/6}.$$

14.3 Determine the posterior probability density $p_X(x|y)$ if the conditional probability density is given by

$$p_Y(y|x) = \frac{1}{\sqrt{2\pi}} e^{-(y-x)^2/2},$$

the prior probability density is given by

$$p_X(x) = \frac{1}{\sqrt{2\pi 4}} e^{-x^2/8},$$

and the marginal probability density is given by

$$p_Y(y) = \frac{1}{\sqrt{2\pi 5}} e^{-y^2/10} .$$

14.4 Consider the real, independent, Gaussian random variables X and Y with identical variances σ^2 and means x_0 and y_0, respectively. Find the PDF $p_Z(z; z_0, \sigma_z^2)$ for the complex random variable Z with mean z_0 such that $Z = X + iY$.

14.5 Consider a random column vector of length N whose elements are drawn independently from a complex, zero-mean circularly symmetric Gaussian distribution of unit variance. Identify and explain the PDF for this random vector.

14.6 Consider the conditional probability for y given x:

$$p_Y(y|x) = \frac{1}{\sqrt{2\pi 4}} e^{-(y-x)^2/8} .$$

Given this relationship, what is the maximum likelihood estimate for x if the value of y is observed to be y_0. Explain your answer.

14.7 For the PDF of the real random variable X,

$$p_X(x) = \frac{1}{\sqrt{2\pi}} e^{-x^2/2} ,$$

(a) evaluate the PDF for $p_Y(y)$ if $y = a(x - b)$, where $a, b \in \mathbb{R}$;
(b) evaluate the variance of Y;
(c) evaluate the skew of Y; and
(d) evaluate the excess kurtosis of Y.

14.8 Evaluate the mean, variance, skew, and excess kurtosis of $p_X(x)$ given by

$$p_X(x) = \begin{cases} x/8 & ; & 0 \le x \le 4 \\ 0 & ; & \text{otherwise} . \end{cases}$$

14.9 Consider the real random variable x that has a PDF given by $p_X(x) = 1$ for $\|x\| < 1/2$ and $f(x) = 0$, otherwise. Evaluate the PDF of y such that $y = ax$, where $a \in \mathbb{R}$ and $a > 0$.

14.10 Consider a complex random variable n drawn from the complex circularly symmetric Gaussian distribution with unit variance. Identify the PDF of the magnitude of n, which is given by $q = \|n\|$. Evaluate the probability that the magnitude q exceeds some threshold η.

15 Fourier Analysis

In this chapter we discuss Fourier analysis. We categorize signals into energy or power signals. We introduce the foundational concept of the complex tone. We define the Fourier transform and identify it as a linear operator. We review energy and the power spectral densities. We survey a set of useful Fourier transform relationships such as time shift, frequency shift, scaling, time reversal, and conjugation. We evaluate the Fourier transform of the top hat function, real Gaussian function, convolution, functional derivative, and autocorrelation. We introduce the Fourier series and evaluate series coefficients for sawtooth functions and impulse train. We discuss the discrete-time Fourier transform and the discrete Fourier transform and related fast Fourier transform. Finally, we review digital filters.

Given the infinite set of possible transforms, it must say something fundamental about either us or the universe that we spend so much effort focused on the Fourier transform. There are innumerable transforms that are available mathematically; however, we are particularly drawn to the Fourier transform. It may be because we, humans, are pretty good at spectral analysis. Based upon the structure of the harmonics that we hear, we can immediately classify instruments among many playing the same note. In engineering, viewing signals in both the spectral and temporal domains provides many insights, and this is particularly true for communications.

15.1 Energy versus Power Signals

When discussing Fourier analysis of signals, we first must consider if the signals have finite energy. As a practical matter, any signal we produce has finite energy; however, in theoretical evaluations, signals that have finite power but exist for all time are common. These signals have infinite energy and may, consequently, have some technical issues. To be technically concrete, some signal $x(t)$ is considered to be a finite-energy signal (or just an energy signal) if

$$\lim_{T\to\infty} \int_{-T/2}^{T/2} dt\, \|x(t)\|^2 = E, \tag{15.1}$$

where E exists (that is, it is some finite number). We have a power signal if

$$\lim_{T\to\infty} \frac{1}{T} \int_{-T/2}^{T/2} dt\, \|x(t)\|^2 = P, \tag{15.2}$$

where $0 < P < \infty$.

For many analyses, we assume a unit resistive load. Often, we are comparing energies or powers to other energies or powers, so the absolute scale is not important. If we do need the actual value, we can scale by the true impedance.

15.2   Complex Tone

The common definition for frequency is the rate of periodic occurrence; however, we have a more precise definition as engineers. In discussing frequency, both "real" frequency f in units of hertz (which is per second) and angular frequency ω in units of radians per second are useful. The two forms are related by

$$\omega = 2\pi f. \tag{15.3}$$

Because the "real" frequency is that used by test equipment, we will typically employ it, but the annoyance of the extra 2π in many equations can be avoided by using angular frequency.

We can define a complex tone by considering a unit length vector rotating at a constant angular rate with frequency f in the complex plane. As a function of time t, the tone $s(t)$ is defined by

$$s(t) = e^{i\,2\pi f t}$$
$$= \cos(2\pi f t) + i\,\sin(2\pi f t). \tag{15.4}$$

This tone moving through time is depicted in Figure 15.1. A complex tone with frequency f_0 has a spectral representation indicated in Figure 15.2, where the Dirac delta is represented with the "lollipop" symbol.

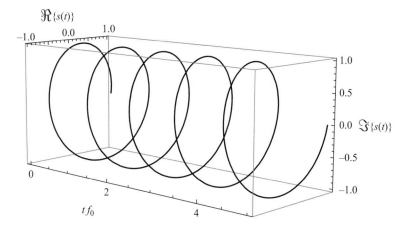

Figure 15.1 A segment of a complex tone $s(t)$ with real frequency f_0 as a function of time t.

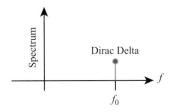

Figure 15.2 Notional spectral representation of the complex tone of frequency f_0 as a function of real frequency f.

15.3   Fourier Transform Definition

Joseph Fourier's transform [40] can be considered the answer to the following question: How much is the signal like a tone at frequency f? Mathematically, this question is asked by taking the function inner product between some general function $g(t)$ and the function that represents our tone $s(t;f)$, where we have extended the notation to be explicit about the use of f. By defining an inner product between complex functions $\langle h(x), g(x) \rangle$ as

$$\langle h(x), g(x) \rangle = \int_{-\infty}^{\infty} dx \, h^*(x) \, g(x), \tag{15.5}$$

this inner product for our complex tone reference is expressed as

$$\langle s(t;f), g(t) \rangle = \int_{-\infty}^{\infty} dt \, s^*(t;f) \, g(t)$$

$$= \int_{-\infty}^{\infty} dt \, [e^{i2\pi f t}]^* \, g(t)$$

$$= \int_{-\infty}^{\infty} dt \, e^{-i2\pi f t} \, g(t) \tag{15.6}$$

$$= \mathcal{F}\{g(t)\} = G(f),$$

where $G(f)$ is the Fourier transform of the function of $g(t)$. The inverse Fourier transform is given by

$$g(t) = \mathcal{F}^{-1}\{G(f)\}$$

$$= \int_{-\infty}^{\infty} df \, e^{i2\pi f t} \, G(f). \tag{15.7}$$

The inverse Fourier transform of the Fourier transform of a function is the original function, such that

$$g(t) = \mathcal{F}^{-1}\{\mathcal{F}\{g(t)\}\}. \tag{15.8}$$

The Fourier transform can also be expressed in terms of angular frequency, which is given by

$$\mathcal{F}\{g(t)\} = G(\omega)$$

$$= \int_{-\infty}^{\infty} dt \, e^{-i\omega t} \, g(t), \tag{15.9}$$

and the inverse transform is given by

$$\mathcal{F}^{-1}\{G(\omega)\} = g(t)$$
$$= \frac{1}{2\pi} \int_{-\infty}^{\infty} d\omega\, e^{i\omega t}\, G(\omega). \tag{15.10}$$

The asymmetry of the constant between the transform and inverse transform is aesthetically displeasing, so a symmetric or unitary form is sometimes employed, such that

$$\tilde{\mathcal{F}}\{g(t)\} = \tilde{G}(\omega)$$
$$= \frac{1}{\sqrt{2\pi}} \int_{-\infty}^{\infty} dt\, e^{-i\omega t}\, g(t), \tag{15.11}$$

and the inverse transform is given by

$$\tilde{\mathcal{F}}^{-1}\{\tilde{G}(\omega)\} = g(t)$$
$$= \frac{1}{\sqrt{2\pi}} \int_{-\infty}^{\infty} d\omega\, e^{i\omega t}\, \tilde{G}(\omega). \tag{15.12}$$

For most of the discussions within this text, we will employ the real frequency version of the Fourier transform because it makes a direct connection with the frequency described in a laboratory setting.

The Fourier transform of a complex tone produces a delta function $\delta(x)$, which is a function of unit area but infinite height at $x = 0$. This is expected from our motivation for our definition of the Fourier transform. For the complex tone $s(t; f_0) = e^{i 2\pi f_0 t}$, the Fourier transform is given by

$$\mathcal{F}\{s(t; f_0)\} = \int_{-\infty}^{\infty} dt\, e^{-i 2\pi f t}\, s(t; f_0)$$
$$= \int_{-\infty}^{\infty} dt\, e^{-i 2\pi f t}\, e^{i 2\pi f_0 t}$$
$$= \int_{-\infty}^{\infty} dt\, e^{i 2\pi (f_0 - f) t} = \delta([f_0 - f] t). \tag{15.13}$$

15.3.1 Trigonometric Expansion

While the exponential form of the Fourier transform, as defined in Equation (15.6), is typically the most useful form, it is sometimes of value to consider the trigonometric expansion of the exponential tone. The Fourier transform can be expressed in terms of sine and cosine transforms, as given by

$$\mathcal{F}\{s(t)\} = \int_{-\infty}^{\infty} dt\, e^{-i 2\pi f t}\, s(t)$$
$$= \int_{-\infty}^{\infty} dt\, [\cos(2\pi f t) - i \sin(2\pi f t)]\, s(t)$$
$$= \int_{-\infty}^{\infty} dt\, \cos(2\pi f t)\, s(t) - i \int_{-\infty}^{\infty} dt\, \sin(2\pi f t)\, s(t). \tag{15.14}$$

15.3.2 Energy or Power Spectral Density

The Fourier transform of a signal is a complex function. Often we want to consider the power or energy spectral density (ESD) of the signal. For a finite-energy signal, the ESD $E(f)$ of the time-domain signal $x(t)$ is given by the magnitude squared of the Fourier transform of the time-domain signal,

$$E(f) = \|X(f)\|^2 . \tag{15.15}$$

For a power signal, we consider the power spectral density (PSD) $P(f)$ because it is not typically useful to think about functions that are infinite for a large range of values. The PSD is given by

$$P(f) = \lim_{T \to \infty} \frac{1}{T} \left\| \int_{-T/2}^{T/2} dt \, e^{-i2\pi f t} x(t) \right\|^2 . \tag{15.16}$$

15.3.3 Linear Operator

The Fourier transform (as well as the inverse transform) is a linear operation. We can see this directly by noting that it satisfies both scaling and summation properties of operators. The scaling property is demonstrated by

$$\mathcal{F}\{a\,x(t)\} = \int_{-\infty}^{\infty} dt \, e^{-i2\pi f t} a\,x(t)$$

$$= a \int_{-\infty}^{\infty} dt \, e^{-i2\pi f t} x(t)$$

$$= a \, \mathcal{F}\{x(t)\} . \tag{15.17}$$

Similarly, the summation property is shown by

$$\mathcal{F}\{x(t) + y(t)\} = \int_{-\infty}^{\infty} dt \, e^{-i2\pi f t} [x(t) + y(t)]$$

$$= \int_{-\infty}^{\infty} dt \, e^{-i2\pi f t} x(t) + \int_{-\infty}^{\infty} dt \, e^{-i2\pi f t} y(t)$$

$$= \mathcal{F}\{x(t)\} + \mathcal{F}\{y(t)\} . \tag{15.18}$$

15.4 Fourier Transform Relationships

Here, we consider a few interesting relationships involving Fourier transforms of the complex function $x(t) = x_{\text{re}}(t) + i\,x_{\text{im}}(t)$, where $x_{\text{re}}(t)$ and $x_{\text{im}}(t)$ indicate the real and imaginary components of the function $x(t)$, respectively.

15.4.1 Time Shift

For our complex signal $x(t)$ with Fourier transform $X(f)$, consider shifting the signal to a later time by amount τ. In the spectral domain, the shift in time corresponds to a phase rotation, given by

$$\mathcal{F}\{x(t-\tau)\} = \int_{-\infty}^{\infty} dt \, e^{-i2\pi f t} \, x(t-\tau)$$

$$\text{by using} \quad t' = t - \tau$$

$$= \int_{-\infty}^{\infty} dt' \, e^{-i2\pi f (t'+\tau)} \, x(t')$$

$$= \int_{-\infty}^{\infty} dt' \, e^{-i2\pi f t'} \, e^{-i2\pi f \tau} \, x(t')$$

$$= e^{-i2\pi f \tau} \int_{-\infty}^{\infty} dt' \, e^{-i2\pi f t'} \, x(t')$$

$$= e^{-i2\pi f \tau} X(f). \tag{15.19}$$

Here we have assumed that our function is well-behaved at infinite time (positive and negative) such that a finite shift in the limits has no effect. For any time-domain signal of finite temporal extent, this is valid. The phase rotation modifies the Fourier transform of the undelayed spectral-domain function by a complex exponential that is a function of the frequency and the delay.

15.4.2   Frequency Shift

The evaluation of a shift in frequency of f_c to larger values (to the right), as displayed in Figure 15.3, is formally similar to the shift in time. This transformation is particularly important to communications because we typically operate at complex baseband (which is near zero frequency or, equivalently, near DC that we discuss in Section 15.4.6), but need to transmit a signal at some higher carrier frequency. The process to convert between baseband and passband is often called up-conversion, which we discussed in Chapter 5. The frequency shift discussed here is part of that process. For the frequency-domain signal $X(f)$ that is shifted by f_c, the time-domain signal is modified by a complex exponential, which is given by

$$\mathcal{F}^{-1}\{X(f-f_c)\} = \int_{-\infty}^{\infty} df \, e^{i2\pi f t} \, X(f-f_c)$$

$$\text{by using} \quad f' = f - f_c$$

$$= \int_{-\infty}^{\infty} df' \, e^{i2\pi (f'+f_c) t} \, X(f')$$

$$= e^{i2\pi f_c t} \int_{-\infty}^{\infty} df' \, e^{i2\pi f' t} \, X(f')$$

$$= e^{i2\pi f_c t} \, x(t). \tag{15.20}$$

The phase rotation modifies the time-domain signal with an exponential function that is a function of the frequency shift. This exponential function is sometimes called a phase ramp because of the linear shift in phase as a function of time. From the perspective of up-conversion, only real signals can be transmitted, so we typically transmit the real component of this complex up-conversion.

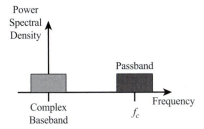

Figure 15.3 Notional spectral shift from around DC to some higher frequency f_c.

15.4.3   Temporal Scaling

The Fourier transform $X(f)$ of a signal $x(t)$ for which time is scaled by a positive-definite value a is given by

$$\mathcal{F}\{x(at)\} = \int_{-\infty}^{\infty} dt\, e^{-i2\pi ft} x(at)$$

by using $\tau = at$

$$= \int_{\infty}^{-\infty} \frac{d\tau}{a}\, e^{-i2\pi f\tau/a} x(\tau)$$

$$= \int_{\infty}^{-\infty} \frac{d\tau}{a}\, e^{-i2\pi f/a\tau} x(\tau)$$

$$= \frac{1}{a} X(f/a). \tag{15.21}$$

15.4.4   Time Reversal

Consider the Fourier transform of the time reversal of signal $x(t)$ given by $x(-t)$. The Fourier transform of the time-reversed signal is given by

$$\mathcal{F}\{x(-t)\} = \int_{-\infty}^{\infty} dt\, e^{-i2\pi ft} x(-t)$$

by using $\tau = -t$

$$= \int_{\infty}^{-\infty} (-d\tau)\, e^{i2\pi f\tau} x(\tau)$$

$$= \int_{-\infty}^{\infty} d\tau\, e^{i2\pi f\tau} x(\tau)$$

$$= \int_{-\infty}^{\infty} d\tau\, e^{-i2\pi f'\tau} x(\tau); \quad f' = -f$$

$$= X(f') = X(-f). \tag{15.22}$$

Consequently, reversing time reverses the spectral shape.

15.4.5   Conjugation

Consider the Fourier transform of a conjugation of the signal $x(t)$ given by $x^*(t)$. The Fourier transform of the conjugated signal is given by

$$x^*(t) = \left(\mathcal{F}^{-1}\{X(f)\} \right)^*$$

$$= \left(\int_{-\infty}^{\infty} dt\, e^{i2\pi f t} X(f) \right)^*$$

$$= \int_{-\infty}^{\infty} dt\, e^{-i2\pi f t} X^*(f)$$

$$= \int_{-\infty}^{\infty} dt\, e^{i2\pi f' t} X^*(-f')$$

$$\mathcal{F}\{x^*(t)\} = X^*(-f). \tag{15.23}$$

Consequently, conjugation of the time-domain signal conjugates and reverses the spectral signal.

15.4.6   DC

The notation DC comes from the field of direct current power delivery (versus alternating current). In this context, it indicates a signal that does not vary as a function of time. The Fourier transform of a DC (or constant) function with complex value a is given by

$$X(f) = \int_{-\infty}^{\infty} dt\, e^{-i2\pi f t} a$$

$$= a\,\delta(f). \tag{15.24}$$

The spectrum is infinite at zero frequency, as shown in Figure 15.4.

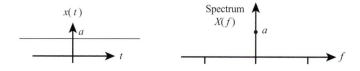

Figure 15.4   Fourier transform $X(f)$ of notional DC signal $x(t)$ with constant amplitude a.

15.4.7   Sines and Cosines

First, let us consider the Fourier transform of the cosine. For some cosine signal of complex amplitude a and frequency f_0, the Fourier transform is given by

$$x(t) = a\,\cos(2\pi f_0 t) \tag{15.25}$$

$$X(f) = \int_{-\infty}^{\infty} dt\, e^{-i2\pi f t} a\,\cos(2\pi f_0 t)$$

$$= \int_{-\infty}^{\infty} dt\, e^{-i2\pi f t}\, a \left(\frac{e^{i2\pi f_0 t} + e^{-i2\pi f_0 t}}{2} \right)$$

$$= \frac{a}{2}\left[\delta(f - f_0) + \delta(f + f_0) \right], \tag{15.26}$$

as is notionally displayed in Figure 15.5. Similarly, the Fourier transform of a sine signal is given by

$$x(t) = a\, \sin(2\pi f_0 t) \tag{15.27}$$

$$X(f) = \int_{-\infty}^{\infty} dt\, e^{-i2\pi f t}\, a\, \sin(2\pi f_0 t)$$

$$= \int_{-\infty}^{\infty} dt\, e^{-i2\pi f t}\, a \left(\frac{e^{i2\pi f_0 t} - e^{-i2\pi f_0 t}}{2i} \right)$$

$$= \frac{a}{2i}\left[\delta(f - f_0) - \delta(f + f_0) \right]$$

$$= \frac{ia}{2}\left[\delta(f + f_0) - \delta(f - f_0) \right]. \tag{15.28}$$

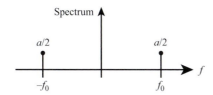

Figure 15.5 Fourier transform of cosine signal with complex amplitude a and frequency f_0.

15.4.8 Odd and Even Functional Symmetries

Any function $x(t)$ can be expressed in terms of the sum of even and odd functions, such that

$$x(t) = x_{\text{even}}(t) + x_{\text{odd}}(t) \tag{15.29}$$

$$x_{\text{even}}(-t) = x_{\text{even}}(t) \tag{15.30}$$

$$x_{\text{odd}}(-t) = -x_{\text{odd}}(t). \tag{15.31}$$

By using Euler's formula, presented in Equation (1.4), we can decompose the exponential form of the Fourier transform into a sum of sine and cosine forms. We simplify the resulting form by noting that the integral over odd integrands evaluates to zero, such that

$$X(f) = \int_{-\infty}^{\infty} dt\, e^{-i2\pi f t}\, x(t) \tag{15.32}$$

$$= \int_{-\infty}^{\infty} dt\, [\cos(2\pi f t) - i\sin(2\pi f t)]\, [x_{\text{even}}(t) + x_{\text{odd}}(t)]$$

$$= \int_{-\infty}^{\infty} dt\, \cos(2\pi f t)\, x_{\text{even}}(t) - i \int_{-\infty}^{\infty} dt\, \sin(2\pi f t)\, x_{\text{odd}}(t). \tag{15.33}$$

We have used the observation that the product of two even functions is an even function, the product of two odd functions is an even function, and the product of an even function and an odd function is an odd function.

15.4.9   Top Hat

Two extremely important functions in communications, which are related by the Fourier transform, are the top hat and the related sinc function. A top hat of unit area with width T and height $1/T$ is defined by

$$x(t) = \begin{cases} \frac{1}{T} & ; & \|t\| \leq \frac{T}{2} \\ 0 & ; & \text{otherwise} \end{cases} , \tag{15.34}$$

as depicted in Figure 15.6. The Fourier transform of this function is given by

$$
\begin{aligned}
X(f) &= \int_{-\infty}^{\infty} dt \, e^{-i 2\pi f t} \, x(t) \\
&= \int_{-T/2}^{T/2} dt \, e^{-i 2\pi f t} \frac{1}{T} \\
&= \frac{1}{T} \frac{1}{-i 2\pi f} \, e^{-i 2\pi f t} \Big|_{-T/2}^{T/2} \\
&= \frac{1}{T} \frac{1}{-i 2\pi f} \left(e^{-i 2\pi f T/2} - e^{-i 2\pi f (-T/2)} \right) \\
&= \frac{1}{\pi f T} \sin(\pi f T) = \text{sinc}(f T) , \tag{15.35}
\end{aligned}
$$

which is depicted in Figure 15.7, where we have employed the standard signal processing definition of the sinc function.

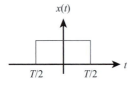

Figure 15.6 Top hat function of width T and height $1/T$ that is centered at $t = 0$.

15.4.10   Real Gaussian

The Fourier transform of a Gaussian-shaped signal is proportional to a Gaussian. We define the time-domain signal $x(t)$ to have a real Gaussian shape that is centered on zero with variance σ_t^2, such that

$$x(t) = \frac{1}{\sqrt{2\pi \, \sigma_t^2}} \, e^{-t^2/[2\sigma_t^2]} . \tag{15.36}$$

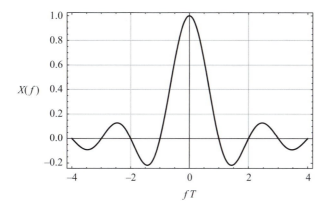

Figure 15.7 Sinc function resulting from Fourier transform of the top hat function defined in Equation (15.35).

The Fourier transform of this function is given by

$$X(f) = \int_{-\infty}^{\infty} dt\, e^{-i2\pi f t}\, x(t)$$

$$= \int_{-\infty}^{\infty} dt\, e^{-i2\pi f t}\, \frac{1}{\sqrt{2\pi\,\sigma_t^2}}\, e^{-\frac{t^2}{2\sigma_t^2}}$$

$$= \frac{1}{\sqrt{2\pi\,\sigma_t^2}} \int_{-\infty}^{\infty} dt\, \exp\left\{-\left(i2\pi f t + \frac{t^2}{2\sigma_t^2} + a^2 - a^2\right)\right\}$$

$$= \frac{1}{\sqrt{2\pi\,\sigma_t^2}}\, \exp\{a^2\} \int_{-\infty}^{\infty} dt\, \exp\left\{-\left(i2\pi f t + \frac{t^2}{2\sigma_t^2} + a^2\right)\right\}$$

$$a^2 = -2\pi^2 f^2 \sigma_t^2 \tag{15.37}$$

$$= \frac{1}{\sqrt{2\pi\,\sigma_t^2}}\, e^{-2\pi^2 f^2 \sigma_t^2} \int_{-\infty}^{\infty} dt\, \exp\left\{-\left(i\sqrt{2\pi} f \sigma + \frac{t}{\sqrt{2}\sigma_t}\right)^2\right\}$$

$$\frac{t'}{\sqrt{2}\sigma_t} = \frac{t}{\sqrt{2}\sigma_t} + i\sqrt{2\pi} f \sigma_t$$

$$= e^{-2\pi^2 f^2 \sigma_t^2}\, \frac{1}{\sqrt{2\pi\,\sigma_t^2}} \int_{-\infty}^{\infty} dt'\, e^{-\frac{t'^2}{2\sigma_t^2}}$$

$$= e^{-2\pi^2 f^2 \sigma_t^2} = \sqrt{2\pi\,\sigma_f^2}\left(\frac{1}{\sqrt{2\pi\,\sigma_f^2}}\, e^{-f^2/(2\sigma_f^2)}\right) \tag{15.38}$$

$$\sigma_f^2 = \frac{1}{(2\pi\,\sigma_t)^2}. \tag{15.39}$$

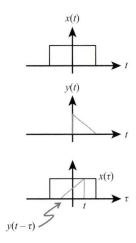

Figure 15.8 Example of convolving function $x(t)$ with $y(t)$.

15.4.11   Convolution

The convolution of two functions $x(t)$ and $y(t)$ is denoted by $x(t) * y(t)$, or sometimes $[x * y](t)$, with the former being more common even though the latter makes more sense formally. The convolution of $x(t)$ and $y(t)$ is given by

$$[x * y](t) = \int_{-\infty}^{\infty} d\tau\, x(\tau)\, y(t - \tau)$$

$$= \int_{-\infty}^{\infty} d\tau\, x(t - \tau)\, y(\tau). \tag{15.40}$$

A graphical example of the convolution of two functions is presented in Figure 15.8. The Fourier transform of the convolution is given by

$$\mathcal{F}\{[x * y](t)\} = \int_{-\infty}^{\infty} dt\, e^{-i2\pi f t} \int_{-\infty}^{\infty} d\tau\, x(\tau)\, y(t - \tau)$$

$$= \int_{-\infty}^{\infty} d\tau\, x(\tau) \int_{-\infty}^{\infty} dt\, e^{-i2\pi f t}\, y(t - \tau)$$

$$= \int_{-\infty}^{\infty} d\tau\, x(\tau) \int_{-\infty}^{\infty} dt\, e^{-i2\pi f (t' + \tau)}\, y(t'); \quad t' = t - \tau$$

$$= \left(\int_{-\infty}^{\infty} d\tau\, e^{-i2\pi f \tau} x(\tau) \right) \left(\int_{-\infty}^{\infty} dt\, e^{-i2\pi f t'}\, y(t') \right)$$

$$= X(f)\, Y(f), \tag{15.41}$$

where $X(f) = \mathcal{F}\{x(t)\}$ and $Y(f) = \mathcal{F}\{y(t)\}$. Thus, we have the wonderfully useful result that the Fourier transform of the convolution of two functions is the product of the Fourier transforms of the individual functions.

15.4.12 Derivative

The Fourier transform of the derivative of a function has a simple form that multiplies the Fourier transform of the original function by the frequency and a constant. First, we rewrite the derivative chain rule as

$$\frac{\partial}{\partial t}[g(t)\,x(t)] = \left(\frac{\partial}{\partial t}g(t)\right)x(t) + g(t)\left(\frac{\partial}{\partial t}x(t)\right)$$

$$g(t)\left(\frac{\partial}{\partial t}x(t)\right) = \frac{\partial}{\partial t}[g(t)\,x(t)] - \left(\frac{\partial}{\partial t}g(t)\right)x(t). \tag{15.42}$$

By applying this form, the transform is given by

$$\mathcal{F}\left\{\frac{\partial}{\partial t}x(t)\right\} = \int_{-\infty}^{\infty} dt\, e^{-i2\pi ft}\,\frac{\partial}{\partial t}x(t)$$

$$= \int_{-\infty}^{\infty} dt\,\frac{\partial}{\partial t}[e^{-i2\pi ft}\,x(t)] - \int_{-\infty}^{\infty} dt\left(\frac{\partial}{\partial t}e^{-i2\pi ft}\right)x(t)$$

$$= e^{-i2\pi ft}\,x(t)\Big|_{-\infty}^{\infty} - \int_{-\infty}^{\infty} dt\,(-i2\pi f)\,e^{-i2\pi ft}\,x(t)$$

$$= 0 + i2\pi f\,\mathcal{F}\{x(t)\}, \tag{15.43}$$

where we have assumed that $x(t)$ goes to zero as t goes to either $\pm\infty$.

15.4.13 Autocorrelation and Spectral Density

The ESD of a finite energy signal is given by the Fourier transform of the autocorrelation.[1] At first, this is somewhat surprising because we usually have to square the Fourier transform to express the energy density; however, the autocorrelation is already sort of a squared quantity. In general, the autocorrelation is a function of the absolute time offsets of the two functions; however, we will assume that we are considering a wide-sense stationary random process (see Section 14.2.3), so we will only address the relative delay. The autocorrelation provides insight into how much a signal at some time is like the signal at some later time. The finite-energy signal autocorrelation is given by

$$R_x(\tau) = \int_{-\infty}^{\infty} dt\, x(t)\,x^*(t - \tau) \tag{15.44}$$

$$= \int_{-\infty}^{\infty} dt\, x(t + \tau)\,x^*(t). \tag{15.45}$$

The two forms are equivalent because a finite shift in time will have no effect on the integral because we are integrating from $-\infty$ to ∞.

---

[1] The autocovariance and autocorrelation calculations are nearly identical. In the case of the autocovariance, the means of the signals are removed. Because we typically use zero-mean signals, there would be no difference.

To simplify development, we assume that the mean of our signal is zero: $E\{x(t)\} = 0$. The Fourier transform of the autocorrelation is given by

$$
\begin{aligned}
\mathcal{F}\{R_x(\tau)\} &= \int_{-\infty}^{\infty} d\tau \, e^{-i2\pi f\tau} \int_{-\infty}^{\infty} dt \, x(t) \, x^*(t-\tau) \\
&= \int_{-\infty}^{\infty} dt \, x(t) \int_{-\infty}^{\infty} d\tau \, e^{-i2\pi f\tau} \, x^*(t-\tau) \\
&= \int_{-\infty}^{\infty} dt \, x(t) \int_{-\infty}^{\infty} d\tau' \, e^{-i2\pi f(t-\tau')} \, x^*(\tau'); \qquad \tau = t - \tau' \\
&= \int_{-\infty}^{\infty} dt \, x(t) \, e^{-i2\pi ft} \int_{-\infty}^{\infty} d\tau' \, x^*(\tau') \, e^{i2\pi f\tau'} \\
&= \left\| \int_{-\infty}^{\infty} dt \, x(t) \, e^{-i2\pi ft} \right\|^2 = E(f),
\end{aligned}
\tag{15.46}
$$

where $E(f)$ is the ESD.

For power signals, we wish to evaluate the PSD. The form for the autocorrelation is not bounded for $\tau = 0$, so the form in Equation (15.45) is not viable. Consequently, we substitute a normalized version such that the autocorrelation is $r_x(\tau)$. In communications, we often think of our signal $x(t)$ as a random process, and we can express this normalized autocorrelation $r_x(\tau)$ as the expectation of $x(t) x^*(t-\tau)$ or, equivalently, $x(t+\tau) x^*(t)$, which is given by

$$
r_x(\tau) = E\{x(t) x^*(t-\tau)\} = E\{x(t+\tau) x^*(t)\}.
\tag{15.47}
$$

The Fourier transform of $r_x(\tau)$ is the PSD $P(f)$:

$$
P(f) = \mathcal{F}\{r_x(\tau)\}.
\tag{15.48}
$$

If the process is ergodic, which indicates that averaging over time provides the same result as the expectations, then the autocorrelation is given by

$$
r_x(\tau) = \lim_{T\to\infty} \frac{1}{T} \int_{-T/2}^{T/2} dt \, x(t) \, x^*(t-\tau).
\tag{15.49}
$$

Again, we have assumed $E[x(t)] = 0$.

15.5　Fourier Series Definition

We assume that we have a signal that repeats with some periodicity T, such that

$$
x(t) = x(t - mT) \,\forall\, m \in \mathbb{Z}.
\tag{15.50}
$$

In this case, the time-domain function $x(t)$ can be represented by the series representation

$$
x(t) = \sum_{m=-\infty}^{\infty} c_m \, e^{i2\pi mt/T}
\tag{15.51}
$$

$$
c_m = \frac{1}{T} \int_{-T/2}^{T/2} dt \, e^{-i2\pi mt/T} \, x(t).
\tag{15.52}
$$

What is notable in this notation is that a continuous time-domain signal can be represented with a discrete (if potentially infinite) series of tones $e^{i 2\pi m t/T}$ in the spectral domain.

It might be surprising that the function can be represented with a set of discrete frequencies as opposed to a continuous frequency-domain form that we see for the general Fourier transform. We can motivate this frequency constraint with the following observation. The general Fourier transform for $x(t)$ is given by

$$X(f) = \int_{-\infty}^{\infty} dt\, e^{-i 2\pi f t}\, x(t)\,; \tag{15.53}$$

however, this must also be true if we shift t by mT because of Equation (15.50). Consequently, we have the relationship

$$
\begin{aligned}
X(f) &= \int_{-\infty}^{\infty} dt\, e^{-i 2\pi f t}\, x(t - mT) \\
&= \int_{-\infty}^{\infty} dt'\, e^{-i 2\pi f (t'+mT)}\, x(t') \quad;\, t' = t - mT \\
&= e^{-i 2\pi f mT} \int_{-\infty}^{\infty} dt\, e^{-i 2\pi f t'}\, x(t') \\
&= e^{-i 2\pi f mT}\, X(f)\,.
\end{aligned}
\tag{15.54}
$$

This relationship is satisfied if $e^{-i 2\pi f mT} = 1$ or, equivalently, if the product of f and T is an integer, which is satisfied if

$$f = \frac{n}{T} \,\forall\, n \in \mathbb{Z}\,, \tag{15.55}$$

that is, f is some multiple of $(1/T)$.

As an explicit example, we consider the sawtooth signal defined by

$$x(t) = t\,; \quad -1/2 < t \le 1/2\,, \tag{15.56}$$

and repeats so that $x(t - m) = x(t)\,\forall\, m \in \mathbb{Z}$. We evaluate the Fourier series coefficients, given by

$$
\begin{aligned}
c_m &= \int_{-1/2}^{1/2} dt\, e^{-i 2\pi m t/T}\, t \\
&= \frac{i\,[\pi\, m\, \cos(\pi\, m) - \sin(\pi\, m)]}{2\,\pi^2\, m^2}\,.
\end{aligned}
\tag{15.57}
$$

By using Equation (15.51), we construct an approximation by only using terms up to some N such that

$$x(t) \approx \sum_{m=-N}^{N} c_m\, e^{i 2\pi m t/T}\,. \tag{15.58}$$

For the sawtooth signal example, we see that the Fourier series reconstruction of the signal converges relatively to the original form that we depicted in Figure 15.9.

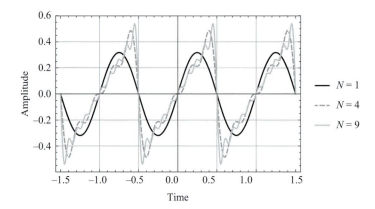

Figure 15.9 Fourier series reconstructed approximation to a sawtooth waveform for terms $m \in \{-N, \cdots, N\}$ with $N = 1, 4, 9$.

15.5.1   Fourier Transform of Impulse Train

An interesting example of Fourier transforms is for the case of a regular train of impulses (or delta functions) as seen in Figure 15.10.

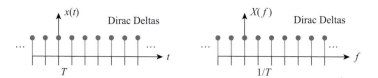

Figure 15.10 Time-domain and frequency-domain representation of a regular train of impulses.

The regular train of unit area impulses is given by

$$x(t) = \sum_{m=-\infty}^{\infty} \delta(t - mT). \tag{15.59}$$

The direct evaluation of the Fourier transform of this "function" is problematic; however, we can evaluate it by representing $x(t)$ with its Fourier series. The series representation is given by

$$x(t) = \sum_{m=-\infty}^{\infty} c_m e^{i 2\pi m t/T} \tag{15.60}$$

$$c_m = \frac{1}{T} \int_{-T/2}^{T/2} dt \, e^{-i 2\pi m t/T} x(t)$$

$$= \frac{1}{T} \int_{-T/2}^{T/2} dt \, e^{-i 2\pi m t/T} \sum_{m=-\infty}^{\infty} \delta(t - mT)$$

$$= \frac{1}{T} \int_{-T/2}^{T/2} dt\, e^{-i2\pi\, mt/T}\, \delta(t) = \frac{1}{T} \tag{15.61}$$

$$x(t) = \sum_{m=-\infty}^{\infty} \frac{1}{T} e^{i2\pi\, mt/T}. \tag{15.62}$$

We can then take the Fourier transform of the Fourier series decomposition of the function. The Fourier transform is given by

$$X(f) = \int_{-\infty}^{\infty} dt\, e^{-i2\pi ft} \left(\sum_{-\infty}^{\infty} \frac{1}{T} e^{i2\pi\, mt/T} \right)$$

$$= \sum_{-\infty}^{\infty} \frac{1}{T} \int_{-\infty}^{\infty} dt\, e^{-i2\pi ft}\, e^{i2\pi\, mt/T}$$

$$= \frac{1}{T} \sum_{-\infty}^{\infty} \int_{-\infty}^{\infty} dt\, e^{i2\pi\, t(m/T-f)}$$

$$= \frac{1}{T} \sum_{-\infty}^{\infty} \delta(f - m/T) \tag{15.63}$$

which is a train of impulses in the frequency domain with area $1/T$ that is spaced at an interval $1/T$.

15.6 Discrete-Time Fourier Transform

If we consider a band-limited spectrum as displayed in Figure 15.11, then we know because of Nyquist that we can exactly reconstruct the signal even if we sample discretely in time. If we have a complex sample spaced at $T = 1/B$, then the Nyquist requirement is satisfied. Note that there is something disturbing in the finite-bandwidth requirement. In general, this implies a signal that has infinite temporal support, which is not actually possible. We circumvent this requirement by satisfying it approximately in real life.

If we sample a finite-energy signal in time at an interval of T and indicate this signal x_m, then the discrete-time Fourier transform (DTFT) is given by

$$X(f) = \sum_{m=-\infty}^{\infty} e^{-i2\pi fTm}\, x_m. \tag{15.64}$$

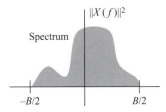

Figure 15.11 Notional spectral representation of a band-limited signal contained within total bandwidth B.

Often we assume units of time such that $T = 1$. We can always convert back to typical units when we are done with an evaluation. In this case, the frequency f is expressed in normalized units such that $-1/2 \leq f < 1/2$. The energy spectral density of the signal is given by

$$E(f) = \|X(f)\|^2 . \qquad (15.65)$$

15.6.1 Periodogram

Given some finite set of N samples of a time-domain signal, it is reasonable to ask what the power spectral density is of this finite-energy signal. In general, we cannot find that answer from only N samples of what must be a longer sequence to satisfy the finite-bandwidth constraint required by Nyquist, but we can make an estimate of the PSD. By using a piece of the DTFT, squaring it, and normalizing it by the number of temporal samples, we find the Schuster's periodogram, which is given by

$$\hat{P}(f) = \frac{1}{N}\|X(f)\|^2$$

$$= \frac{1}{N}\left\|\sum_{m=0}^{N-1} e^{-i\,2\pi f\,m}\,x_m\right\|^2 . \qquad (15.66)$$

We use the $\hat{P}(f)$ to indicate that this is just an estimate. However, if the signal repeats temporally every N samples, then the estimate is the actual PSD. Of course, practically, that does not make any sense.

15.6.2 Discrete Autocorrelation and Power Spectral Density

Similar to our discussion in Section 15.4.13, we define our autocorrelation $r_x(k)$ for a discretely sampled wide-sense stationary random sequence of signals x_m with relative sample delay k by using

$$r_k = E\{x_m\,x_{m-k}^*\} = E\{x_{m+k}\,x_m^*\} . \qquad (15.67)$$

If x_m is deterministic, then we just directly calculate it. An example might be that we are evaluating the autocorrelation of a known sampled channel impulse response. If x_m is drawn from some random data, then we estimate autocorrelation by considering N samples, given by

$$\hat{r}_k = \frac{1}{N}\sum_{m=k+1}^{N} x_m\,x_{m-k}^* ; \quad 0 \geq k \geq N - 1 . \qquad (15.68)$$

We could alternatively use an unbiased estimator, given by

$$\tilde{r}_k = \frac{1}{N-k}\sum_{m=k+1}^{N} x_m\,x_{m-k}^* ; \quad 0 \geq k \geq N - 1 . \qquad (15.69)$$

Because for most communications signals the autocorrelation r_k falls quickly as k grows, for reasonable values of N these two forms provide similar results. The first version is more common. We observe the symmetry

$$r_{-k} = r_k^* . \tag{15.70}$$

By evaluating the discrete-time Fourier transform of the autocorrelation, we get an estimate of the PSD $\hat{P}(f)$ as a function of frequency f, given by

$$\hat{P}(f) = \sum_{k=-(N-1)}^{N-1} \hat{r}_k \, e^{-i 2\pi f k T} , \tag{15.71}$$

where T is the sample period.

We observe that this estimate converges asymptotically to the true PSD. In the following calculation, we use the double sum relationship given by

$$\sum_{m=1}^{N} \sum_{n=1}^{N} g(m - n) = \sum_{k=-(N-1)}^{N-1} [N - \|k\|] \, g(k) , \tag{15.72}$$

for some arbitrary function $g(k)$. We recover the PSD $P(f)$ from the magnitude squared of the DTFT in the limit of infinite number of samples, given by

$$P(f) = \lim_{N\to\infty} \frac{1}{N} E\left\{ \left\| \sum_{m=1}^{N} x_m \, e^{-i 2\pi f m T} \right\|^2 \right\} \tag{15.73}$$

$$= \lim_{N\to\infty} \frac{1}{N} E\left\{ \sum_{m=1}^{N} \sum_{n=1}^{N} x_m \, e^{-i 2\pi f m T} x_n^* \, e^{i 2\pi f n T} \right\}$$

$$= \lim_{N\to\infty} \frac{1}{N} \sum_{m=1}^{N} \sum_{n=1}^{N} E[x_m x_n^*] \, e^{-i 2\pi f (m-n) T}$$

$$= \lim_{N\to\infty} \frac{1}{N} \sum_{k=-(N-1)}^{N-1} [N - \|k\|] \, r_k \, e^{-i 2\pi f k T}$$

$$= \lim_{N\to\infty} \sum_{k=-(N-1)}^{N-1} r_k \, e^{-i 2\pi f k T} - \frac{1}{N} \sum_{k=-(N-1)}^{N-1} \|k\| \, r_k \, e^{-i 2\pi f k T} . \tag{15.74}$$

For typical signals, the second term in Equation (15.74) goes to zero in the limit of large N because the autocorrelation becomes small for large k.

15.7 Discrete Fourier Transform

The discrete Fourier transform (DFT) is one of the most commonly used tools in signal processing, particularly in its fast Fourier transform (FFT) implementation. Amusingly, the requirements for it to be technically valid are essentially never satisfied, although the results are approximately correct. The implication of using this approach is that estimates are biased. We first define the DFT and then warn about its potential misuse.

If we define a sampled signal sequence in time by an N-dimensional column vector $\mathbf{x} \in \mathbb{C}^{N\times 1}$, for which the mth element is indicated by x_m, and the DFT of this sequence to be the column vector $\mathbf{y} \in \mathbb{C}^{N\times 1}$, for which the mth element is indicated by y_m, then the DFT is given by

$$y_k = \sum_{m=0}^{N-1} e^{-i 2\pi \, \Delta f \, k \, m \, T} \, x_m,$$

(15.75)

where we have assigned $f = \Delta f \, k$ and Δf is the spacing of the frequency samples. Because the resolution in frequency of a signal of duration $N\,T$ is $1/(N\,T)$, the natural value of Δf is $\Delta f = 1/(N\,T)$. The DFT is then given by

$$y_k = \sum_{m=0}^{N-1} e^{-i 2\pi \, k \, m/N} \, x_m.$$

(15.76)

The inverse DFT is given by

$$x_k = \frac{1}{N} \sum_{m=0}^{N-1} e^{i 2\pi \, k \, m/N} \, y_m.$$

(15.77)

This transformation can be written succinctly using the form

$$\mathbf{y} = \mathbf{F}\,\mathbf{x},$$

(15.78)

where the discrete Fourier transformation matrix is defined by

$$\mathbf{F} = \begin{pmatrix} \alpha^{0 \cdot 0} & \alpha^{0 \cdot 1} & \alpha^{0 \cdot 2} & \cdots & \alpha^{0 \cdot (N-1)} \\ \alpha^{1 \cdot 0} & \alpha^{1 \cdot 1} & \alpha^{1 \cdot 2} & \cdots & \alpha^{1 \cdot (N-1)} \\ \vdots & & & \ddots & \\ \alpha^{(N-1) \cdot 0} & \alpha^{(N-1) \cdot 1} & \alpha^{(N-1) \cdot 2} & \cdots & \alpha^{(N-1) \cdot (N-1)} \end{pmatrix}.$$

(15.79)

We have defined the constant α as

$$\alpha = e^{-i 2\pi/N}.$$

(15.80)

The inverse transform is given by

$$\mathbf{x} = \mathbf{F}^{-1}\,\mathbf{y}.$$

(15.81)

Interestingly, the DFT matrix $\mathbf{F}$ is almost unitary (up to a factor of $\sqrt{N}$), so that the inverse Fourier transform matrix is given by

$$\begin{aligned} \mathbf{F}^{-1} &= \frac{1}{N}\mathbf{F}^{H} \\ &= \frac{1}{N} \begin{pmatrix} \alpha^{0 \cdot 0} & \alpha^{0 \cdot 1} & \alpha^{0 \cdot 2} & \cdots & \alpha^{0 \cdot (N-1)} \\ \alpha^{1 \cdot 0} & \alpha^{1 \cdot 1} & \alpha^{1 \cdot 2} & \cdots & \alpha^{1 \cdot (N-1)} \\ \vdots & & & \ddots & \\ \alpha^{(N-1) \cdot 0} & \alpha^{(N-1) \cdot 1} & \alpha^{(N-1) \cdot 2} & \cdots & \alpha^{(N-1) \cdot (N-1)} \end{pmatrix}^{H} \\ &= \frac{1}{N} \begin{pmatrix} \beta^{0 \cdot 0} & \beta^{0 \cdot 1} & \beta^{0 \cdot 2} & \cdots & \beta^{0 \cdot (N-1)} \\ \beta^{1 \cdot 0} & \beta^{1 \cdot 1} & \beta^{1 \cdot 2} & \cdots & \beta^{1 \cdot (N-1)} \\ \vdots & & & \ddots & \\ \beta^{(N-1) \cdot 0} & \beta^{(N-1) \cdot 1} & \beta^{(N-1) \cdot 2} & \cdots & \beta^{(N-1) \cdot (N-1)} \end{pmatrix}, \end{aligned}$$

(15.82)

where $\beta = \alpha^* = e^{i 2\pi/N}$.

We nearly always misuse the DFT in practice and say that, from an engineering perspective, it is good enough. This is a valid worldview; however, it is worth keeping in mind the implications of our common abuse of the underlying assumptions. The DFT assumes that we can represent a signal in time by both a finite set of regular samples in time and a finite set of regular samples in frequency. In general, if the temporal extent in time is finite, then the bandwidth is infinite. Similarly, if the spectral extent in frequency is finite, then the temporal extent is infinite. We assumed that both were finite in extent. What is worse, we assumed that they can be represented by a finite number of samples. We could get around this by assuming that our N samples in time and frequency are in fact periodic so that they both repeat. We only need the N samples to represent the signal. Because the signal repeats in time, it only has support on a regular lattice of points in frequency. We can use the same argument for repeating in frequency for support on a regular lattice of points in time. Consequently, we have constructed a requirement of the signal that, while it is on a lattice in time and frequency, it has infinite extent because it repeats every N lattice points in either the time or frequency direction. Clearly, this is problematic; however, we often just ignore the underlying assumptions and say that spectral estimates produced by the DFT are good enough. For some applications, we use windowing or tapering techniques to reduce the estimation biases. Discussions of these approaches are beyond the scope of this text.

Because of the structure in the DFT matrix $\mathbf{F}$, the evaluation of the matrix vector product can be accelerated significantly, particularly if N is a power of 2. In practice, it is so common to use FFTs to employ DFTs that FFT is sometimes used when we actually mean DFT. The FFT enables the evaluation of the DFT with order $\mathcal{O}(n\log(n))$ operations rather than the $\mathcal{O}(n^2)$ of a matrix-vector multiple observed in Equation (15.78). This potential computational saving motivates a number of signal processing and communications techniques. As an example, orthogonal frequency-division multiplexing (OFDM), discussed in Section 7.5, typically exploits this computational saving.

15.8 Filtering

As we introduced in Section 2.2.5, we need filters to remove out-of-band noise or interference. Conceptually, digital and analog filtering are related. Digital filtering combines delayed and scaled copies of the signal to produce the desired spectral effect. For analog circuits, there are multiple approaches. One can construct filters from the appropriate connection of inductors, capacitors, and resistors. Alternatively, electromechanical devices can be employed, such as surface acoustic wave (SAW) filters that exploit mechanical wave propagation via piezoelectric couplings to a crystal or ceramic. The filtering is produced by combining multiple paths.

Digital filters enable spectral shaping in the digital domain, often applied at complex baseband. Often filters are categorized as either finite-impulse-response (FIR) or infinite-impulse-response (IIR). The FIR filter has memory of the input of the filter.

The IIR filter has memory of the filter output. Consequently, the IIR filter can potentially have an infinite sequence of nonzero outputs even if there is only one nonzero input (an impulse). In general, filters may have both FIR and IIR contributions.

If the nth complex sample input to a filter is indicated by x_n, then the kth filter output y_k is described by

$$y_k = \sum_{m=0}^{p} b_m x_{m-k} - \sum_{m=1}^{q} a_m y_{m-k}, \tag{15.83}$$

where the filter coefficients a_m and b_m correspond to the IIR and FIR terms, respectively. It is sometimes useful to consider the filter in terms of its Z-transform, indicated by $\mathcal{Z}\{\cdot\}$. Each unit delay is represented by the term z^{-1}. By reorganizing the terms in Equation (15.83), we have

$$y_k + \sum_{m=1}^{q} a_m y_{m-k} = \sum_{m=0}^{p} b_m x_{m-k}$$

$$\sum_{m=0}^{q} a_m y_{m-k} = \sum_{m=0}^{p} b_m x_{m-k}; \quad a_0 = 1$$

$$\mathcal{Z}\left\{\sum_{m=0}^{q} a_m y_{m-k}\right\} = \mathcal{Z}\left\{\sum_{m=0}^{p} b_m x_{m-k}\right\}$$

$$A(z)\,Y(z) = B(z)\,X(z), \tag{15.84}$$

where $X(z)$ and $Y(z)$ indicate the polynomials that represent the filter input and output sequences, respectively, and $A(z)$ and $B(z)$ indicate the polynomials that represent the filter coefficients. Here, we have used the observation that the Z-transform of the convolution of two sequences is the product of the individual sequence Z-transforms. Explicitly, $A(z)$ and $B(z)$ are given by

$$A(z) = \sum_{m=0}^{q} a_m z^{-m} \tag{15.85}$$

$$B(z) = \sum_{m=0}^{p} b_m z^{-m}. \tag{15.86}$$

For the filter to be stable (not run off to infinity for finite inputs), the zeros of the polynomial $A(z) = 0$ must lie within the unit circle: $\|z\| < 1$. The transfer function $H(z)$ that indicates the effect of the filter is given by

$$H(z) = \frac{B(z)}{A(z)}$$

$$= \frac{\sum_{m=0}^{p} b_m z^{-m}}{\sum_{m=0}^{q} a_m z^{-m}}. \tag{15.87}$$

The PSD response $P(f)$ of the filter under the assumption of a unit-variance spectrally white input signal is given by

$$P(f) = \|H(z)\|^2 \text{ with } z = e^{i 2\pi f}$$
$$= \left\| H(e^{i 2\pi f}) \right\|^2, \tag{15.88}$$

where the f is the normalized frequency limited by $-1/2 \leq f < 1/2$.

As an example, we use the Parks–McClellan optimal equiripple FIR filter design approach to build a filter. The details of this filter design approach are beyond the scope of this text, but to illustrate, the filter coefficients of this 9-tap filter are given by

$$\mathbf{b} = \begin{pmatrix} 0.0046 \\ -0.0624 \\ -0.0399 \\ 0.1668 \\ 0.4104 \\ 0.4104 \\ 0.1668 \\ -0.0399 \\ -0.0624 \\ 0.0046 \end{pmatrix}. \tag{15.89}$$

The filter shape for $\mathbf{b}$ is depicted in Figure 15.12. For the example, we have a spectral passband region of $\|f\| < 0.2$ in normalized frequency and in the out-of-band region $\|f\| > 0.3$, the signal is suppressed by about 30 dB.

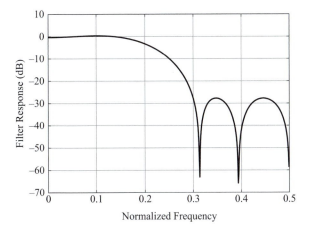

Figure 15.12 Filter response of FIR filter presented in Equation (15.89).

Problems

15.1 Construct the time-domain representation $s(t)$ for a signal that contains three complex tones with frequencies f_1, f_2, and f_3 and amplitudes a_1, a_2, and a_3, respectively.

15.2 Explain whether the following waveforms are energy signals or power signals:
(a) $s(t) = e^{-t^2}$
(b) $s(t) = \sin(10\pi t)$
(c) $s(t) = \max\{0, 3 - \|t\|\}$
(d) $s(t) = 1 + i$
(e) $s(t) = \frac{1}{t-i}$

15.3 Evaluate the Fourier transform of a top hat function of unit height and width.

15.4 Evaluate the Fourier transform of $a \cos(2\pi f_c t)$ with respect to t.

15.5 Evaluate the Fourier transform of $\frac{\partial}{\partial t} a \sin(2\pi f_c t)$ with respect to t.

15.6 Evaluate the inverse Fourier transform of the triangle waveform $S(f) = 1 - \|f\|$ for $\|f\| \leq 1$.

15.7 Evaluate the Fourier transform of $s(t) = \cos(2\pi t)$ for $\|t\| < 1/4$ and $s(t) = 0$ otherwise.

15.8 Evaluate the Fourier series coefficients of a periodic linear ramp of duration T and extreme value of $-1/2$ and $1/2$.

15.9 Evaluate the Fourier series coefficients of a periodic signal given by $s(t) = t^2 - 1/12$ for $\|t\| \leq 1/2$ and repeats for $s(t - m) \in \mathbb{Z}$.

15.10 Consider the DFT.
(a) Write the 4×4 DFT matrix $\mathbf{F}$ explicitly in terms of complex exponentials.
(b) Evaluate $\mathbf{F} (0 \quad 0 \quad 1 \quad 0)^T$.
(c) Evaluate $\mathbf{F} (1 \quad 1 \quad 1 \quad 1)^T$.
(d) Evaluate $\mathbf{F} (1 \quad e^{-i 2\pi 2/4} \quad e^{-i 2\pi 4/4} \quad e^{-i 2\pi 6/4})^T$.

15.11 Evaluate the DFT of the vectors
(a) $\{1, 1, 1, 1, 1, 1, 1, 1\}^T$
(b) $\{1, 0, 0, 0, 0, 0, 0, 0\}$
(c) $\{0, 0, 0, 1, 0, 0, 0, 0\}$

15.12 Use a simulation to:
(a) evaluate the DFT of a vector of dimension 64 where the entries are drawn independently from a zero-mean unit-variance real Gaussian distribution;
(b) use the DFT evaluation to determine the ESD; and
(c) repeat ESD evaluation and plot the spectral density evaluation averaging within a frequency bin for 1, 10, and 100 draws.

15.13 For a signal proportional to the top hat function with unit height and width T,

(a) evaluate the ESD by directly considering the Fourier transform of the top hat; and

(b) evaluate the energy density by employing the autocorrelation of the top hat.

15.14 By evaluating the autocorrelation of $s(t) = 1/12 - t^2$ for $\|t\| < 1/2$ and 0 otherwise, find the ESD.

15.15 If $s(t)$ is a real odd function, explain what is known about the Fourier transform.

15.16 Evaluate the convolution of $\cos(2\pi f_c t)$ and $\delta(t - 1/2)$.

15.17 For the triangle waveform $s(t) = 1 - \|t\|$ for $\|t\| \leq 1$ and zero otherwise, evaluate the ESD.

15.18 Evaluate the Fourier series coefficients for the waveform $s(t) = \cos(\pi t)$ for $\|t\| \leq 1/2$ that repeats so that $s(t - m) = s(t)$ for $m \in \mathbb{Z}$.

15.19 Consider the function defined by

$$x(t) = (t/T)^2 \quad \forall \, \|t\| \leq T/2$$
$$x(t - mT) = x(t) \quad \forall \, m \in \mathbb{Z}.$$

(a) Evaluate the Fourier series coefficients for $x(t)$.

(b) By using MATLAB® or equivalent, plot and compare the function $x(t)$ with the approximation of the function $x(t)$ built from the Fourier series using coefficients $m \in \{-3, \cdots, 3\}$.

15.20 Show that for well-behaved functions $x(t)$ and $y(t)$

$$\int_{-\infty}^{\infty} dt \, x(t) \, y^*(t) = \int_{-\infty}^{\infty} df \, X(f) \, Y^*(f) \, .$$

15.21 Show that the Fourier transform of the convolution of two functions $x(t)$ and $y(t)$ is the product of the Fourier transform of functions $X(f)$ and $Y(f)$

$$\mathcal{F}\{[x * y](t)\} = X(f) \, Y(f) \, .$$

15.22 Evaluate the Fourier transform of a uniformly spaced temporal sequence of identical impulses.

15.23 Consider a channel that is characterized by a tap delay line with coefficient values of 10^{-6} at zero delay and $-0.9 \cdot 10^{-6}$ at one-sample delay.

(a) Evaluate the Z-transform of the transfer function.

(b) Evaluate the filtering effect on PSD of the channel as a function of frequency.

(c) Plot the PSD of the filter on a decibel scale.

References

[1] Bhagwandas P. Lathi and Zhi Ding. *Modern Digital and Analog Communication Systems*, 5th edition. Oxford University Press, 2019.

[2] John G. Proakis and Masoud Salehi. *Communication Systems Engineering*, 2nd edition. Pearson Education, 2002.

[3] Upamanyu Madhow. *Introduction to Communication Systems*. Oxford University Press, 2014.

[4] Mary L. Boas. *Mathematical Methods in the Physical Sciences*. John Wiley & Sons, 2006.

[5] Constantine A. Balanis. *Modern Antenna Handbook*. John Wiley & Sons, 2011.

[6] H. Nyquist. Thermal agitation of electric charge in conductors. *Physical Review*, 32:110–113, July 1928.

[7] C. E. Shannon. A mathematical theory of communication. *Bell System Technical Journal*, 27:379–423, July 1948.

[8] C. E. Shannon. Communication in the presence of noise. *Proceedings of the IRE*, 37(1):10–21, Jan. 1949.

[9] T. M. Cover and J. A. Thomas. *Elements of Information Theory*, 2nd edition. John Wiley & Sons, 2006.

[10] John G. Proakis. *Digital Communications*. McGraw-Hill, 2001.

[11] Norbert Wiener. *Extrapolation, Interpolation, and Smoothing of Stationary Time Series*. John Wiley & Sons, 1949.

[12] John G. Proakis and Dimitris G. Manolakis. *Digital Signal Processing*. Pearson Prentice Hall, 2007.

[13] R. W. Chang. Synthesis of band-limited orthogonal signals for multichannel data transmission. *The Bell System Technical Journal*, 45(10):1775–1796, Dec. 1966.

[14] M. Hata. Empirical formula for propagation loss in land mobile radio services. *IEEE Transactions on Vehicular Technology*, 29(3):317–325, Aug. 1980.

[15] R. W. Hamming. Notes on digital coding. *Bell System Technical Journal*, 29:147–160, Apr. 1950.

[16] A. Viterbi. Error bounds for convolutional codes and an asymptotically optimum decoding algorithm. *IEEE Transactions on Information Theory*, 13(2):260–269, Apr. 1967.

[17] Shu Lin and Daniel J. Costello. *Error Control Coding*. Prentice Hall, 2005.

[18] William E. Ryan and Shu Lin. *Channel Codes: Classical and Modern*. Cambridge University Press, 2009.

[19] Milton Abramowitz and Irene A. Stegun. *Handbook of Mathematical Functions: With Formulas, Graphs, and Mathematical Tables*. Dover Publications, 2012.

[20] A. Sabharwal, P. Schniter, D. Guo, D. W. Bliss, S. Rangarajan, and R. Wichman. In-band full-duplex wireless: challenges and opportunities. *IEEE Journal on Selected Areas in Communications*, 32(9):1637–1652, Sept. 2014.

[21] Daniel W. Bliss and Siddhartan Govindasamy. *Adaptive Wireless Communications: MIMO Channels and Networks*. Cambridge University Press, 2013.

[22] V. Tarokh, N. Seshadri, and A. R. Calderbank. Space–time codes for high data rate wireless communication: performance criterion and code construction. *IEEE Transactions on Information Theory*, 44(2):744–765, Mar. 1998.

[23] V. Tarokh, H. Jafarkhani, and A. R. Calderbank. Space–time block codes from orthogonal designs. *IEEE Transactions on Information Theory*, 45(5):1456–1467, July 1999.

[24] S. M. Alamouti. A simple transmit diversity technique for wireless communications. *IEEE Journal in Selected Areas in Communications*, 16:1451–1458, Oct. 1998.

[25] Andrew J. Viterbi. *Principles of Coherent Communication*. McGraw-Hill, 1966.

[26] Erwin Kreyszig. *Advanced Engineering Mathematics*. John Wiley & Sons, 2006.

[27] George F. Carrier, Max Krook, and Carl E. Pearson. *Functions of a Complex Variable: Theory and Technique*. Society for Industrial and Applied Mathematics, 2005.

[28] Gene Howard Golub and Charles F. Van Loan. *Matrix Computations*. Johns Hopkins University Press, 1996.

[29] Dennis S. Bernstein. *Matrix Mathematics: Theory, Facts, and Formulas*. Princeton University Press, 2009.

[30] Kenneth S. Miller. *Some Eclectic Matrix Theory*. Robert E. Krieger Publishing, 1987.

[31] A. Graham. *Kronecker Products and Matrix Calculus*. Ellis Horwood Limited, 1981.

[32] Robert A. Monzingo, Randy L. Haupt, and Thomas W. Miller. *Introduction to Adaptive Arrays*, 2nd edition. Scitech Publishing, Inc., 2011.

[33] I. S. Gradshteyn and I. M. Ryzhik. *Table of Integrals, Series, and Products*. Academic Press, 1994.

[34] E. T. Whittaker and G. N. Watson. *A Course of Modern Analysis*. Cambridge University Press, 1927.

[35] George F. Carrier, Max Krook, and Carl E. Pearson. *Functions of Complex Variable: Theory and Technique*. Hod Books, 1983.

[36] S. Stein. Unified analysis of certain coherent and noncoherent binary communications systems. *IEEE Transactions on Information Theory*, 10(1):43–51, Jan. 1964.

[37] Mischa Schwartz, William R. Bennett, and Seymour Stein. *Communication Systems and Techniques*. McGraw-Hill, 1966.

[38] A. Nuttall. Some integrals involving the q_m function. *IEEE Transactions on Information Theory*, 21(1):95–96, Jan. 1975.

[39] E. J. Kelly and K. W. Forsythe. Adaptive Detection and Parameter Estimation for Multidimensional Signal Models. Technical Report 848, MIT Lincoln Laboratory, Apr. 1989.

[40] Ivor Grattan-Guiness and J. R. Ravetz. *Joseph Fourier, 1768–1830: A Survey of His Life and Work*. MIT Press, 2003.

Index